Study on Physicochemical Properties of Food Protein

Study on Physicochemical Properties of Food Protein

Editors

Shudong He
Wei Xu
Muhammad H. Aludatt

MDPI

Basel • Beijing • Wuhan • Barcelona • Belgrade • Novi Sad • Cluj • Manchester

Editors
Shudong He
Department of Food Science and Engineering
Hefei University of Technology
Hefei
China

Wei Xu
Department of Food Science
Wuhan Polytechnic University
Wuhan
China

Muhammad H. Aludatt
Department of Nutrition and Food Technology
Jordan University of Science and Technology
Irbid
Jordan

Editorial Office
MDPI
St. Alban-Anlage 66
4052 Basel, Switzerland

This is a reprint of articles from the Special Issue published online in the open access journal *Molecules* (ISSN 1420-3049) (available at: www.mdpi.com/journal/molecules/special_issues/food_protein_properties).

For citation purposes, cite each article independently as indicated on the article page online and as indicated below:

Lastname, A.A.; Lastname, B.B. Article Title. *Journal Name* **Year**, *Volume Number*, Page Range.

ISBN 978-3-0365-9868-0 (Hbk)
ISBN 978-3-0365-9867-3 (PDF)
doi.org/10.3390/books978-3-0365-9867-3

Contents

molecules MDPI

Editorial

Study on Physicochemical Properties of Food Protein

Shudong He

School of Food Science and Biological Engineering, Hefei University of Technology, Hefei 230009, China; shudong.he@hfut.edu.cn

1. Introduction

As the global population continues to grow, the demand for sustainable and nutritious food sources has never been higher. Edible proteins, derived from both plant and animal sources, play a crucial role in providing essential amino acids and contributing to the overall quality of our foods. Thus, the food industry has witnessed a surge of interest in understanding and utilizing the physicochemical properties of proteins derived from various sources, and the study of food protein properties is vital, which in turn, can lead to the development of food products with improved nutritional profiles, enhanced functionalities, and extended shelf lives [1]. Then, the solubility, viscosity, water/oil-holding capacity, gelatinization, and emulsifying and foaming properties of the novel nutritional proteins were widely explored regarding their potential roles in the presentation of texture, flavor, color, and nutrition of foods [2].

Furthermore, recent studies found that the functionality of unfolded or hydrolyzed proteins was different from that of unchanged proteins [3], which revealed the structure–function relationship of proteins. From a structural perspective, studies regarding protein conformation, aggregation, molecular weight, isoelectric point, surface hydrophilicity/hydrophobicity and protein binding characteristics are urgently needed to overcome the knowledge gap between protein exploration and application [4]. Actually, the physical and chemical actions during food processing, such as the mechanical functions, thermodynamic actions, pressure effects, pH, and ionic strength, lead to structural changes in the food proteins as the ingredients further influence the functionality and quality of the food products. In recent research, the effects of novel thermal food processing technologies, such as ultra-high temperature short-time (UHT) heating, extrusion, and microwave heating, as well as non-thermal treatments including high hydrostatic pressure, ultrasound, gamma irradiation, and high voltage electrical discharge, on whey and milk proteins have been preferentially investigated, and the physicochemical and structural properties of related proteins are well known [5]. Moreover, high hydrostatic pressure, ohmic heating, pulsed electric fields, and other emerging processing treatments are called "green technologies", as they also bring new opportunities to avoid the excessive use of water, gas emissions, and energy [6]. In order to improve or create new protein processing properties, physical, chemical, and enzymatic treatments and combined processing have been widely applied in protein modifications. The physical modifications, including temperature, pressure, shear, pH, ions, and electrostatics, might induce structural perturbation in the thermodynamic state of the protein. The chemical modifications have involved glycosylation, acetylation, succinylation, phosphorylation, and lipophilization, which will change the net charge of proteins via binding with amino acids. Enzymic modification is characterized by enzyme specificity, hydrolysis degree, and protease action site, and the functional properties of protein hydrolysates can be controlled under controlled proteolysis, while the protein hydrolysates can also be used for further food development [7–9]. Also, in the past few decades, a number of protein allergens have been identified in foods that induce IgE antibody production and the occurrence of food allergies, and a better understanding of

Citation: He, S. Study on Physicochemical Properties of Food Protein. *Molecules* **2023**, *28*, 8145. https://doi.org/10.3390/molecules28248145

Received: 12 December 2023
Accepted: 15 December 2023
Published: 18 December 2023

the protein structures and amino acid sequences of the epitopes is crucial for food safety consumption [10].

In this Special Issue, "Study on Physicochemical Properties of Food Protein," ten outstanding articles were gathered presenting the latest research and innovations in the related field of food science and technology, focusing on the modification, structure, and application of proteins. The ten articles showcase the potential of proteins as functional ingredients in food products, highlighting their impact on flavor, texture, and stability. These studies not only contribute to our fundamental understanding of food proteins but also offer valuable insights for the food industry in terms of food development and process optimization.

2. An Overview of Published Articles

Rice protein is a quality, gluten-free, plant-based protein with high biological functionalities and low allergenicity. Yang et al. (contribution 1) examined the factors contributing to the limited solubility of rice protein, which included the abundance of hydrophobic amino acid residues, disulfide bonds, and intramolecular hydrogen bonds, and raised the corresponding modification strategies, such as glycosylation, interaction between polyphenols and heterologous proteins, phosphate modification, protein deamidation, enzymatic hydrolysis with the assistance of heating, ultrasound, microwave, ultra-high pressure, enzymes, and alkaline conditions. Moreover, the application of modified rice proteins in dairy, meat, and baked food products was illustrated due to their good solubility, emulsifying properties, gelling properties, water-holding capacity, and oil-holding capacity.

Low-frequency ultrasound (20–100 kHz, 1–1000 W/cm^2), a cost-effective non-thermal process, is increasingly employed in the food industry, and the electrical energy is transmuted into vibrational energy, which will generate an acoustic cavitation effect with sudden localized heat and pressure within a liquid setting to promote bond disruption and formation of–diverse covalent and non-covalent linkages, including hydrogen bonds and hydrophobic affinities, with the release of significant energy (10–100 kJ/mol) and physical shearing forces [11]. Liu et al. (contribution 2) explored the potential effects of ultrasonication on the structural integrity and self-assembly behavior of the extracted bone collagen I. The article found that ultrasound ($\leq$200 W, $\leq$15 min) uniformly disintegrated collagen clusters, significantly escalating collagen self-assembly at a reduced yet consistent fibril diameter and decreasing D periodicity, compared to untreated samples; simultaneously, ultrasound ($\leq$200 W, $\leq$15 min) diminished viscoelasticity index and gel strength, heightening thermal stability while fostering larger specific surface areas and increased porosity in collagen fibril gels to manufacture novel collagen biomaterials with desirable performances. In contribution 3, ultrasonic (240 W) and transglutaminase (20 U/g egg white protein) treatments were applied to destroy the globular structure of egg white protein, which has an abundance of hydrophobic amino acids and could be a potential emulsifier after modification, and an egg white protein W1/O emulsion was created using 4 wt.% egg white protein gel particles diluted in 40% (v/v) soya oil. Then, 0.6% (v/v) of chitosan solutions (W2) was incorporated into this Pickering emulsion to yield an egg white protein–chitosan bilayer emulsion via high-speed homogenization, which would benefit eco-friendly techniques in egg processing, packaging, and distribution.

Recent studies on this topic have researched flavor, color, and the bioactive properties' promoting effects of Maillard reaction products (MRPs) derived from protein hydrolysates, comprising a blend of peptides and amino acids [12]. The impact of ultrasound on the characteristics of MRPs derived from hydrolyzed soybean meal was studied in contribution 4. Prior to the Maillard reaction, the Maillard reaction system, with the addition of cysteine (2.0%), xylose (3.0%), VB1 (0.05%), and enzymatically hydrolyzed lard (1.0%) into the soybean meal enzymatic hydrolysis, was sonicated at 300 W (5 s on; 5 s off) for 30 min. The article revealed that ultrasonic processing escalated unsaturated fatty acid production within the aqueous phase during the lard hydrolysis and had boosting effects on Maillard reaction intermediates and melanoid accumulation in the product, and then both the quantity and flavor-enhancing components of volatiles were enhanced after the ultrasound while the excellent antioxidant features were demonstrated.

In particular, the application of limited hydrolysis and ultrafiltration strategies has exhibited promising results in enhancing the physico-functional attributes of food proteins. In contribution 5, donkey-hide gelatin was treated by the Alcalase 2.4L, and the structural characteristics and antioxidant activities of five molecular weight (MW) hydrolysates, including >30, 10–30, 3–10, 1–3, and <1 kDa fragments, were evaluated. The peptide of MW 1–3 kDa presented the strongest antioxidant activity, which was attributed to the amino acid composition. Furthermore, the amino residues of Arg415, Gly462, Phe478, and Tyr572 were confirmed as the key sites to bind with the Keap1 protein via molecular docking, indicating good potential for antioxidant capacity.

The food processing procedure combining multiple mechanical forces has attracted a lot of attention as it is highly efficient in protein modification. Xu et al. (contribution 6) utilized ultra-high-pressure jet processing on skimmed bovine milk. This approach altered the secondary and tertiary casein structures, contributing to a decrease in the mean particle size and zeta potential of casein micelles. Consequently, flatter, looser, and porous casein micellar morphologies improved curd production efficiency and the texture of fermented milk. In contribution 7, steam flash explosion (SFE) was applied to modify high-temperature denatured defatted rice bran protein isolate (RBPI), and the surface hydrophobicity and thermal stability decreased with an increase in the intrinsic viscosity resulting from the protein unfolding. Then, high-temperature rice bran proteins could be utilized in the food industry with their enhanced solubility.

In contribution 8, the hot-pressed peanut meal was used to produce the plant protein-based adhesive via sodium dodecyl sulfate (SDS) denaturation and polyamide epichlorohydrin (PAE) crosslinking. Papain hydrolysis contributed towards enhancing wet shear strength, whereas SDS and urea disrupted hydrogen bonds for better water resistance. Significantly, PAE amplified both the wet shear strength and water resistance of the adhesive with an increase in viscosity.

In the study presented in contribution 9, esterification was applied to soy proteins, leading to a notable increase in the isoelectric point (pI) of the esterified proteins. This increase was attributed to the formation of esters between alcohol and carboxyl groups, which subsequently reduced the content of negatively charged groups on the protein surface. Consequently, the esterified soy protein nano-emulsions demonstrated remarkable storage stability, freeze–thaw stability, and thermal stability. Furthermore, the esterification modification imparted antibacterial properties. These findings highlight the potential of esterified soy proteins as a promising material in various applications, particularly in the food and pharmaceutical industries.

As one of the nine major food allergens identified by the Food and Drug Administration in the USA, it becomes imperative to elucidate the primary allergenic epitopes of cashew nut proteins. Zhang et al. (contribution 10) utilized DNAstar and PyMoL tools to ascertain potential epitopes for Ana o 2 and Ana o 3 proteins. They then applied the phage display methodology to validate the prospective synthetic epitope peptides. It is noteworthy that Ana o 3 might carry a greater risk of provoking an allergic response compared to Ana o 2, which is primarily attributed to the larger number of identifiable epitopes.

3. Conclusions

The ten research contributions presented in this issue collectively demonstrate the significance and versatility of studying the physicochemical properties of food proteins under various modifications. These studies have explored various aspects, including protein solubility, structural integrity, emulsifying properties, and allergenic epitopes, as well as the application of different processing techniques such as ultrasonication, high-pressure processing, and esterification. Advances in "eco-friendly" technologies provide promising opportunities to diminish water consumption, gas emission rates, and energy utilization within the food industry.

The findings from these studies have not only contributed to a deeper understanding of the role of proteins in food products, but they have also provided valuable insights for the food industry in terms of product development, process optimization, and the potential application of novel protein-based materials. For instance, the use of modified rice proteins in dairy, meat, and baked goods; the prospective use of low-frequency ultrasound in protein manipulation; taste enhancement provided by amino acid compounds resulting from the Maillard reaction; the development of plant protein-based adhesives; the potential application of esterified soy proteins; and limited enzymolysis donkey-hide gelatin in the food and pharmaceutical industries. Furthermore, pioneering processing methodologies like ultra-high-pressure jet processing and steam flash explosions exhibit their effectiveness in modifying protein architectures for better functionality. Of particular interest, the analysis of soybean protein esterification uncovers its potential application in nano-emulsion formations with improved durability and antibacterial properties. The recognition of allergenic epitopes in cashew nut proteins accentuates the necessity of comprehending allergenicity to safeguard consumer well-being.

In conclusion, the research presented in this issue showcases the potential of food proteins and their modifications in various applications, emphasizing the need for continued research and innovation in this field. By further exploring the physicochemical properties of food proteins and developing new processing techniques, the food industry can continue to create high-quality, nutritious, and safe products that cater to the diverse needs of consumers.

Conflicts of Interest: The author declares no conflict of interest.

List of Contributions

1. Yang, J.; Meng, D.; Wu, Z.; Chen, J.; Xue, L. Modification and solubility enhancement of rice protein and its application in food processing: A review. *Molecules* **2023**, *28*, 4078. https://doi.org/10.3390/molecules28104078.
2. Liu, H.; Zhang, H.; Wang, K.; Qi, L.; Guo, Y.; Zhang, C.; Xu, Y. Impact of ultrasonication on the self-assembly behavior and gel properties of bovine bone collagen I. *Molecules* **2023**, *28*, 3096. https://doi.org/10.3390/molecules28073096.
3. Pan, Q.; Zhang, Z.; Wu, R.; Guo, D.; Xu, W.; Wang, H.; Yi, Y. Fabrication and characterization of the egg-white protein chitosan double-layer emulsion. *Molecules* **2022**, *27*, 6036. https://doi.org/10.3390/molecules27186036.
4. Ye, Y.; Dai, S.; Zhang, H.; He, S.; Hu, W.; Cao, X.; Wei, Z. Ultrasound-assisted preparation of Maillard reaction products derived from hydrolyzed soybean meal with meaty flavor in an oil-in-water system. *Molecules* **2022**, *27*, 7236. https://doi.org/10.3390/molecules27217236.
5. Liang, R.; Xu, L.; Fan, C.; Cao, L.; Guo, X. Structural characteristics and antioxidant mechanism of donkey-hide gelatin peptides by molecular dynamics simulation. *Molecules* **2023**, *28*, 7975. https://doi.org/10.3390/molecules28247975.
6. Xu, F.; Xue, L.; Ma, Y.; Niu, T.; Zhao, P.; Wu, Z.; Wang, Y. Effects of ultra-high-pressure jet processing on casein structure and curdling properties of skimmed bovine milk. *Molecules* **2023**, *28*, 2396. https://doi.org/10.3390/molecules28052396.
7. Na, Z.; Bi, H.; Wang, Y.; Guo, Y.; Ma, Y. Effect of steam flash-explosion on physicochemical properties and structure of high-temperature denatured defatted rice bran protein isolate. *Molecules* **2023**, *28*, 643. https://doi.org/10.3390/molecules28020643.
8. Qu, Y.; Guo, Q.; Li, T.; Liu, H.; Wang, Q. Effects of Different denaturants on the properties of a hot-pressed peanut meal-based adhesive. *Molecules* **2022**, *27*, 4878. https://doi.org/10.3390/molecules27154878.

9. Wang, T.; Yi, K.; Li, Y.; Wang, H.; Fan, Z.; Jin, H.; Xu, J. Esterified soy proteins with enhanced antibacterial properties for the stabilization of nano-emulsions under acidic conditions. *Molecules* **2023**, *28*, 3078. https://doi.org/10.3390/molecules28073078.
10. Zhang, A.; Zhao, H.; Pei, S.; Chi, Y.; Fan, X.; Liu, J. Identification and structure of epitopes on cashew allergens Ana o 2 and Ana o 3 using phage display. *Molecules* **2023**, *28*, 1880. https://doi.org/10.3390/molecules28041880.

References

1. Aschemann-Witzel, J.; Gantriis, R.F.; Fraga, P.; Perez-Cueto, F.J. Plant-based food and proteinhydrophobicitybusiness perspective: Markets, consumers, and the challenges and opportunities in the future. *Crit. Rev. Food Sci. Nutr.* **2021**, *61*, 3119–3128. [CrossRef] [PubMed]
2. Zhang, Q.; Li, L.; Lan, Q.; Li, M.; Wu, D.; Chen, H.; Liu, Y.; Lin, D.; Qin, W.; Zhang, Z.; et al. Protein glycosylation: A promising way to modify the functional properties and extend the application in food system. *Crit. Rev. Food Sci. Nutr.* **2019**, *59*, 2506–2533. [CrossRef] [PubMed]
3. Barbiroli, A.; Iametti, S.; Bonomi, F. Beta-lactoglobulin as a model food protein: How to promote, prevent, and exploit its unfolding processes. *Molecules* **2022**, *27*, 1131. [CrossRef] [PubMed]
4. van Schaick, G.; Haselberg, R.; Somsen, G.W.; Wuhrer, M.; Domínguez-Vega, E. Studying protein structure and function by native separation–mass spectrometry. *Nat. Rev. Chem.* **2022**, *6*, 215–231. [CrossRef]
5. Fasolin, L.H.; Pereira, R.N.; Pinheiro, A.C.; Martins, J.T.; Andrade, C.C.P.; Ramos, O.L.; Vicente, A.A. Emergent food proteins–Towards sustainability, health and innovation. *Food Res. Int.* **2019**, *125*, 108586. [CrossRef] [PubMed]
6. María, L.P.; Díaz-Reinoso, B.; Giancarlo, C.; Barba, F.J.; Moure, A.; Domínguez, H.; Daniel, F. 3-Green technologies for food processing: Principal considerations. In *Innovative Thermal and Non-Thermal Processing, Bioaccessibility and Bioavailability of Nutrients and Bioactive Compounds*; Barba, F.J., Saraiva, J.M.A., Cravotto, G., Lorenzo, J.M., Eds.; Woodhead Publishing: Sawston, UK, 2019; pp. 55–103.
7. Hoyt, E.A.; Cal, P.M.; Oliveira, B.L.; Bernardes, G.J.L. Contemporary approaches to site-selective protein modification. *Nat. Rev. Chem.* **2019**, *3*, 147–171. [CrossRef]
8. Nasrabadi, M.N.; Doost, A.S.; Mezzenga, R. Modification approaches of plant-based proteins to improve their techno-functionality and use in food products. *Food Hydrocoll.* **2021**, *118*, 106789. [CrossRef]
9. Singh, T.P.; Siddiqi, R.A.; Sogi, D.S. Enzymatic modification of rice bran protein: Impact on structural, antioxidant and functional properties. *LWT* **2021**, *138*, 110648. [CrossRef]
10. Zhou, F.; He, S.; Sun, H.; Wang, Y.; Zhang, Y. Advances in epitope mapping technologies for food protein allergens: A review. *Trends Food Sci. Technol.* **2021**, *107*, 226–239. [CrossRef]
11. Pandiselvam, R.; Aydar, A.Y.k.; Kutlu, N.; Aslam, R.; Sahni, P.; Mitharwal, S.; Gavahian, M.; Kumar, M.; Raposo, A.; Yoo, S. Individual and interactive effect of ultrasound pre-treatment on drying kinetics and biochemical qualities of food: A critical review. *Ultrason. Sonochemistry* **2023**, *92*, 106261. [CrossRef] [PubMed]
12. Fu, Y.; Zhang, Y.; Soladoye, O.P.; Aluko, R.E. Maillard reaction products derived from food protein-derived peptides: Insights into flavor and bioactivity. *Crit. Rev. Food Sci. Nutr.* **2020**, *60*, 3429–3442. [CrossRef] [PubMed]

molecules

MDPI

Review

Modification and Solubility Enhancement of Rice Protein and Its Application in Food Processing: A Review

Jingjing Yang, Dan Meng, Zijian Wu *, Jinyu Chen and Lu Xue

Tianjin Key Laboratory of Food Biotechnology, College of Biotechnology and Food Science, Tianjin University of Commerce, Tianjin 300134, China; 120200310@stu.tjcu.edu.cn (J.Y.)
* Correspondence: wzjian@tjcu.edu.cn

Abstract: Rice protein is a high-quality plant-based protein source that is gluten-free, with high biological value and low allergenicity. However, the low solubility of rice protein not only affects its functional properties such as emulsification, gelling, and water-holding capacity but also greatly limits its applications in the food industry. Therefore, it is crucial to modify and improve the solubility of rice protein. In summary, this article discusses the underlying causes of the low solubility of rice protein, including the presence of high contents of hydrophobic amino acid residues, disulfide bonds, and intermolecular hydrogen bonds. Additionally, it covers the shortcomings of traditional modification methods and the latest compound improvement methods, compares various modification methods, and puts forward the best sustainable, economical, and environmentally friendly method. Finally, this article lists the uses of modified rice protein in dairy, meat, and baked goods, providing a reference for the extensive application of rice protein in the food industry.

Keywords: rice protein; solubility enhancement; food processing; application

Citation: Yang, J.; Meng, D.; Wu, Z.; Chen, J.; Xue, L. Modification and Solubility Enhancement of Rice Protein and Its Application in Food Processing: A Review. *Molecules* **2023**, *28*, 4078. https://doi.org/10.3390/molecules28104078

Academic Editor: Lina Zhang

Received: 21 April 2023
Revised: 4 May 2023
Accepted: 9 May 2023
Published: 13 May 2023

1. Introduction

Rice protein is a high-quality, gluten-free plant-based protein derived from byproducts of rice processing. It contains a comprehensive range of amino acids (including all 20 essential and non-essential amino acids), a well-balanced ratio, and one of the highest biological values of all cereal crops [1]. Additionally, rice protein is characterized by its light color, mild odor, and low allergenicity [2]. Moreover, rice protein hydrolysates have been reported to possess various biological activities, such as antihypertensive, antioxidant, anticancer, and anti-obesity effects [3–6]. However, the poor water solubility of rice protein significantly restricts its applications in the food industry and related fields.

Protein solubility typically refers to the concentration of protein in a saturated aqueous solution in equilibrium with the solid phase (crystalline or amorphous) under external conditions such as pH, temperature, and ionic strength [7]. Water solubility is one of the important thermodynamic properties of proteins, and it is a prerequisite for other functional properties of proteins, such as emulsification, foaming, gelation, and digestibility. It also further determines the texture and stability of processed food products. However, rice protein has a relatively low overall water solubility, and the high content of alkaline-soluble glutelin makes it difficult to dissolve in slightly acidic or neutral solutions, resulting in low utilization efficiency [8]. This greatly affects its applications in the food and related fields. In order to improve the solubility and utilization of rice protein, various physical (ultrasound, high-pressure, and microwave), chemical (glycosylation, phosphorylation, and deamidation), and biological enzyme-based modification methods can be used, and these techniques can also be combined to produce a simple, safe, and efficient modification of rice protein. The modified rice protein thus obtained has great potential for application in food processing [9].

This article provides an overview of recent methods used for modifying and increasing the solubility of rice protein and further explores the application of modified rice protein in food production, such as dairy products, meat products, and baked goods.

2. Analysis of the Water Solubility Characteristics and Reasons for Rice Protein

According to the solubility of each constituent protein, rice protein can be divided into four categories, of which the most abundant (about 80%) is alkali-soluble globulin, which is mainly composed of 30–40 kDa α (acidic) subunit and 17–23 kDa β (basic) subunit. The other categories include 1–5% alcohol-soluble protein, 2–10% salt-soluble globulin, and 2–5% water-soluble albumin. The main subunits are 11~24 kDa, 19–25 kDa, and 10~100 kDa, respectively [9,10]. The properties of the predominant alkali-soluble glutelin determine the overall properties of rice protein.

The low water solubility of rice glutelin is primarily due to the presence of many hydrogen bonds between and within its molecules, which are formed by the amide groups on the glutamine and asparagine residues interacting with the smaller-sized glycine and alanine residues through steric hindrance [8]. Figure 1 shows the primary structure and amino acid residues composition of rice glutelin from three rice sources (*Oryza sativa*, *Oryza sativa* Japonica Group, and *Oryza sativa* Indica Group). It can be found that rice glutelin contains 17.23–18.60% of glutamine and asparagine residues, and the hydrogen bonds formed by these two amino acid residues replace a large number of hydrogen bonds between amide groups and water molecules, thereby reducing the water solubility of rice glutelin [11].

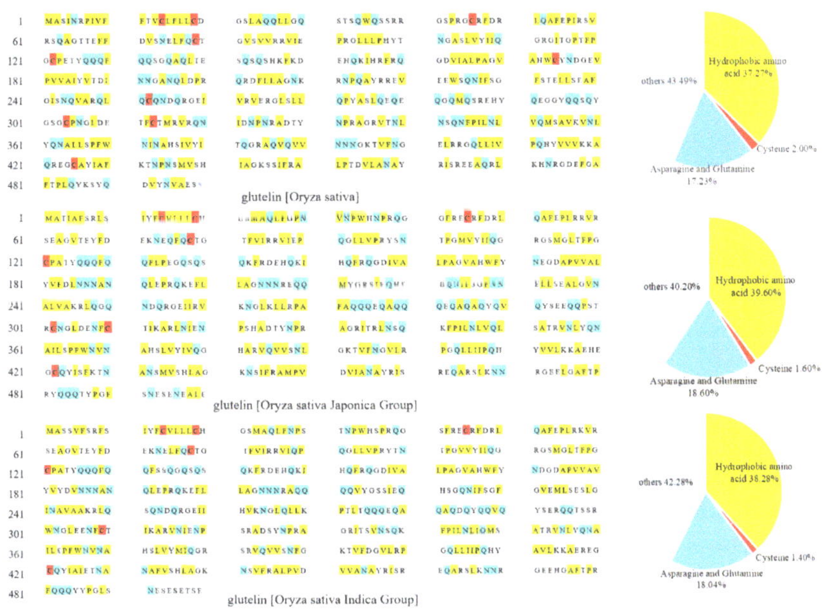

Figure 1. Primary structures and amino acid residue compositions of rice glutelin from three different sources.

Secondly, there are disulfide bonds between rice glutelin subunits, which are formed by the coupling of thiol groups of two different cysteine residues. As shown in Figure 1, rice glutelin contains a small number of cysteine residues (1.40~2.00%), which can interact with each other to generate disulfide bonds within or between subunits, especially between subunits, leading to the formation of large molecular complexes and a decrease in the water solubility of rice glutelin [12].

In addition, rice glutelin contains 37.27~39.60% hydrophobic amino acid residues (as shown in Figure 1), which can cause hydrophobic interactions between amino acid residues and result in the formation of a stable, tensile protein network structure between subunits, which limits the entry of water molecules [13] and therefore reduces its water solubility.

3. Methods for Modifying Rice Protein to Increase Solubility

By taking corresponding measures to address the low water solubility of rice glutelin mentioned above, it is possible to effectively improve the water solubility of rice glutelin and thus enhance the overall solubility of rice protein. The commonly used methods include the following: (1) utilizing the interaction of polysaccharides, polyphenols, composite proteins, and phosphates with rice protein to increase the number of hydrophilic hydroxyl groups or hydrogen bonds formed with water molecules. (2) Using deamidation reactions to break the amide bonds in the side chains of rice protein and reduce the formation of hydrogen bonds between protein subunits. (3) Enzymatically hydrolyzing rice protein to break down the large protein molecules into small peptide molecules and promote their binding to water molecules.

3.1. Modification and Solubility Enhancement of Rice Protein by Polysaccharides

The hydroxyl groups within polysaccharide molecules have strong hydrophilic properties [14]. It is possible to effectively increase the hydrophilic properties of the protein and thus improve its water solubility through covalent or non-covalent interactions between polysaccharides and proteins [15]. For example, the Maillard reaction, in which the carbonyl group of the polysaccharide reacts with the amino group of the protein to form a condensed product [16], can create covalent bonds between protein and polysaccharide molecules. However, the compact globular structure of rice protein makes it challenging to complete the Maillard reaction with polysaccharides in a short period of time [17], thereby reducing the rate and extent of the Maillard reaction and affecting the solubility properties of the conjugated products formed by the traditional Maillard reaction [18]. In recent years, various assisting methods including ultrasound, ultra-high pressure, and microwave have been used to improve or promote the efficiency of the Maillard reaction. These techniques can successfully bring rice protein into adequate contact with polysaccharides, enhancing the reaction efficiency and water solubility. The reaction process is illustrated in Figure 2.

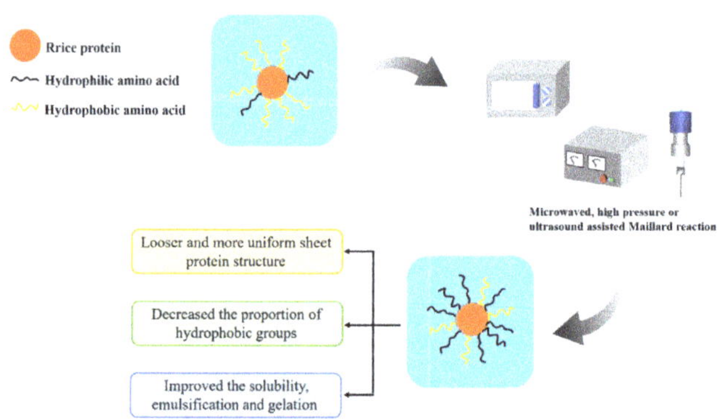

Figure 2. Improvement of solubility of rice protein by using microwaved-, high pressure- or ultrasound-assisted Maillard reaction.

Ultrasound can effectively promote the mixing of protein and polysaccharide molecules, accelerate the penetration speed of solvents into proteins, and thereby increase the Maillard reaction rate [19]. Chen et al. [20] found that after 22 min of ultrasound treatment with a power of 600 W, the spatial conformation of rice protein molecules changed from a globular or blocky structure to a more uniform and loose lamellar structure, which was conducive to the entry of solvents and accelerated the Maillard reaction rate. The solubility of the modified rice protein produced by the process was significantly improved to 90.6%.

Ultra-high-pressure assistance can prevent protein denaturation and aggregation caused by long-term heating, which is more conducive to the effective entry of solvents into protein molecules during the Maillard reaction. Therefore, it is possible to improve the efficiency of the Maillard reaction at lower temperatures [21]. In addition, ultra-high pressure can cause rice protein to expand into a loose spatial conformation, promote the exposure of hydrophobic groups, increase the content of free sulfhydryl groups, and thereby enhance the gel strength of the protein [14]. Xiao et al. [22] prepared rice protein-xylose copolymers with a high grafting degree under ultra-high hydrostatic pressure assistance, which improved the solubility of rice protein (increased from 12.8% to 35.3%) and reduced the formation of harmful by-products such as acrylamide, furan, and hydroxymethyl furfural.

Proper microwave treatment can also shorten the Maillard reaction time and enhance the solubility of rice protein [23]. Meng et al. [24] found that microwave-assisted heating-assisted protein–polysaccharide binding could increase the solubility of rice bran protein to 90.97%. This is because microwaves have the ability to produce strong mechanical collisions and shearing between protein and polysaccharide molecules, which facilitated their adequate interaction. Therefore, more branched glucan may be grafted onto the protein molecules, enhancing the protein's solubility [25].

3.2. Modification and Solubility Enhancement of Rice Protein by Polyphenols

Polyphenols are a group of naturally occurring compounds derived from plants that contain numerous polar hydroxyl groups, including phenolic acids, flavonoids, and tannins, among others [26]. Polyphenols can form non-covalent bonds with proteins, in which the hydrophilic phenolic hydroxyl groups in polyphenols can undergo hydrogen bonding with proteins, and the aromatic rings can interact with the hydrophobic groups of proteins through hydrophobic interactions. Polyphenols can also undergo covalent binding with amino acid residues on proteins, especially lysine residues, cysteine residues, and tryptophan residues, under alkaline conditions, thereby altering the functional properties of the protein [27].

The non-covalent binding between polyphenols and proteins can be simulated by computer docking. Therefore, in this study, AutoDock Vina docking software was used to analyze the interactions between four polyphenols (namely, gallic acid, procyanidins, resveratrol, and ferulic acid), and specific amino acid residues located at the active site of rice glutelin (gene: GLUA2), using its three-dimensional structure as a model (as shown in Figure 3). During the molecular docking process, the polyphenol is considered to be a flexible structure, while the protein is viewed as a rigid structure. The best docking result is chosen based on the principle of minimum docking energy, and Discovery Studio 2021 Client software is used for visualization display. The results showed that the four types of polyphenols mainly formed hydrogen bonds with Thr381, Arg404, Gln110, Gln160, and Gln382, and formed hydrophobic interactions with Val180, Val187, and Arg403. Among them, rice glutelin had more interaction sites with gallic acid, including Asp322, Thr381, Arg404, Gln160, and Gln386, mainly through hydrogen bonding, which was consistent with the research results of Dai [28]. The interaction between rice glutelin and ferulic acid is mainly through hydrophobic interactions, including the interaction with hydrophobic amino acid residues such as Leu407, Ile109, Ile185, Val187, Tyr379, Thr387, Phe430, and Ile428. The interaction between rice glutelin and ferulic acid can lead to a conformational change of rice glutelin, decrease the surface hydrophobicity, and increase water solubility and emulsifying properties [29].

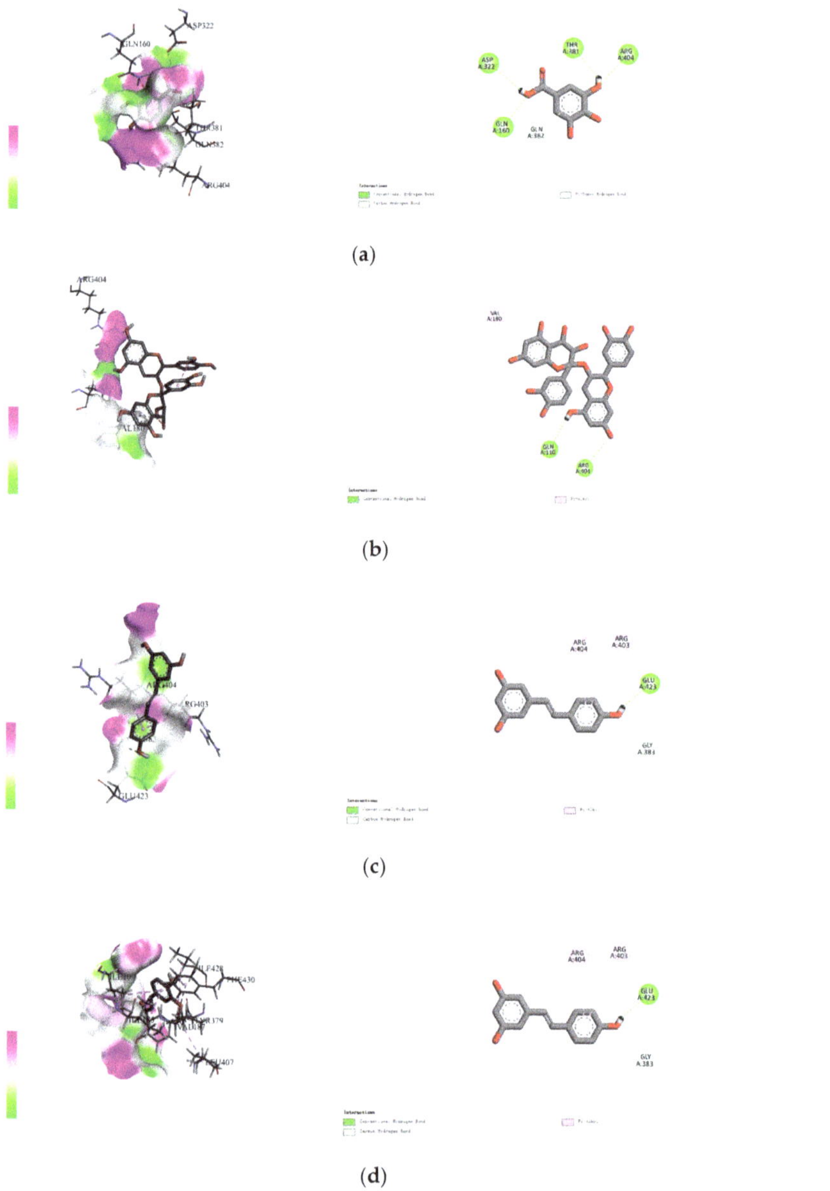

Figure 3. Rice protein–polyphenol molecular docking results. (**a**) gallic acid, (**b**) procyanidins, (**c**) resveratrol, (**d**) ferulic acid.

Alkaline treatment, enzymatic treatment, radical coupling, and ultrasound-assisted methods can further improve the covalent binding efficiency between polyphenols and proteins. According to the alkaline coupling approach, phenolic hydroxyl groups are converted to quinones or semiquinones under alkaline conditions, which can further bind to more amino acid residues and alter the physicochemical properties of the target protein [30]. Wang et al. [29] found that under the condition of pH 9.0, the covalent complex of rice bran protein hydrolysate (10 mg/mL) and ferulic acid (1.5 mg/mL) had strong emulsifying (35.10%) and antioxidant properties (the DPPH clearance rate is 49.70% and the

ABTS+ clearance rate is 89.04%). In addition, protein–polyphenol covalent complexes can also be produced via enzymatic and radical coupling methods [31]. However, traditional methods require long reaction times, which limits large-scale production. Ultrasound-assisted methods can improve the reaction efficiency of enzymatic, alkaline, and radical coupling methods. The underlying principle is that ultrasound treatment can cause protein spatial structure to unfold through mechanical shear, cavitation effects, and thermal effects, increasing the chance of contact reaction between protein and polyphenol molecules, and thus improving covalent reaction efficiency [32]. For example, Xue et al. [33] used ultrasound-assisted radical coupling to improve the grafting efficiency of glutelin and polyphenols, shortened the reaction time, and increased the solubility of glutelin.

3.3. Modification and Solubility Enhancement of Rice Protein by Interaction with Heterologous Proteins

The solubility of rice protein can also be improved by forming a complex with heterologous proteins (usually hydrophilic proteins) through pH cycling. Figure 4 depicts the mechanism of the complex formation using whey protein as an example: first, rice protein and whey protein are mixed at pH 7.0, and then the pH of the solution is adjusted to 12.0 to fully dissolve rice protein, at which point its tertiary structure is completely unfolded and can fully contact with whey protein. The protein solution is then adjusted back to pH 7.0 for protein refolding, and subsequently subjected to centrifugation, dialysis, and freeze-drying to obtain the "rice protein–whey protein" complex [34]. When rice protein and heterologous protein interact to generate a hydrophobic interior, the hydrophilic region is exposed on the surface of the protein, increasing the proportion of surface charge or hydrophilic groups, and ultimately improving the water solubility of rice protein [35].

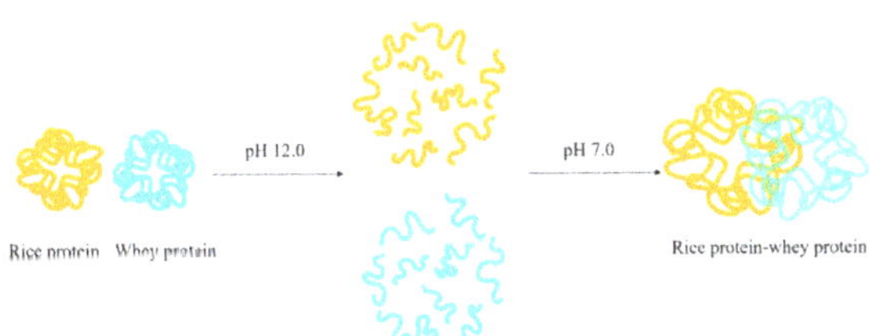

Figure 4. The formation mechanism of the rice protein–whey protein composite.

Wang et al. [36] prepared a composite of rice protein and whey protein in a 1:1 ratio, which resulted in a composite with a solubility of over 50%. The microstructure of the composite showed that the protein molecules were transformed into dispersed, angular, granular structures. Li et al. [37] utilized pH cycling to prepare a composite of rice protein and walnut protein, which increased the solubility of rice protein to over 80%. The hydrophobic groups of the two proteins were buried and included internally, while the charged groups were exposed externally. The resulting composite had an increased zeta potential and reduced surface hydrophobicity. Manhee et al. [38] found that soy protein isolate can also improve the water solubility of rice protein, increasing its solubility from 25.8% to 68.4%.

In addition, the acylation transfer reaction can also be assisted by transglutaminase to cross-link rice glutelin and another heterologous protein [39]. For example, He et al. [40] promoted the cross-linking reaction between rice glutelin and casein by using transglutaminase, which improved the microstructure of rice glutelin and increased its solubility performance.

3.4. Modification and Solubility Enhancement of Rice Protein by Phosphate

Phosphate modification is also a commonly used method to improve the solubility of proteins, and its reaction mechanism is shown in Figure 5. The phosphate group can form hydrogen bonds with the hydroxyl groups on serine, threonine, or tyrosine residues, and increases the number of negative charges, thereby enhancing the electrostatic repulsion between protein molecules and reducing the surface tension of the emulsion, ultimately improving the dispersion of proteins in the emulsion system [41,42]. However, the phosphate modification is currently plagued by the issue of a lengthy reaction time, which can easily cause irreversible aggregation and denaturation of proteins [43]. Therefore, the wet-heat method or microwave method is often used to assist the phosphorylation modification of rice protein to increase the efficiency of the phosphorylation reaction.

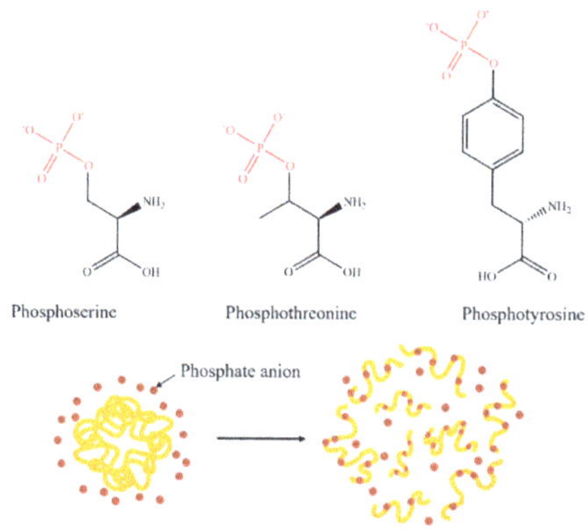

Figure 5. Diagrammatic drawing and structural formula of protein phosphorylation reaction.

Wet-heat-assisted phosphorylation modification involves mixing rice protein with a phosphate solution and incubating it in a 55 °C water bath. This method is effective, simple to operate, and has certain industrial prospects [44]. Hu et al. [45] conducted a wet-heat-assisted phosphorylation reaction on 2% rice bran protein (reaction conditions were 0.1 mol/L sodium trimetaphosphate, pH 9.0, and 55 °C). The conformation of rice bran protein changed, the maximum fluorescence emission wavelength (λ_{max}) shifted to blue, and the surface hydrophobicity and fluorescence intensity were significantly increased. Meanwhile, sodium trimetaphosphate introduced more negative charges on the surface of rice bran protein molecules, increasing their hydrophilicity and the repulsion between protein molecules, thereby improving the solubility of rice bran protein.

Microwave-assisted heating can also improve the efficiency of a phosphorylation reaction. Utilizing the enormous energy generated by electromagnetic waves accelerates the unfolding of protein molecules and the exposure of phosphorylation sites, thereby promoting the interaction between phosphate groups and protein side chains [46,47]. Hadidi et al. [48] used microwave-assisted phosphorylation (at a power of 590 W for 155 s) to increase the degree of protein phosphorylation, shorten the reaction time, and improve the solubility and other functional properties of the protein.

3.5. Modification and Solubility Enhancement of Rice Protein by Deamidation Method

The modification of proteins through deamidation is typically carried out under acidic or alkaline conditions, or enzymatic catalysis, where the amide groups on the side chains of basic amino acid residues (such as asparagine or glutamine residues) are cleaved to form carboxylic acid groups (thus becoming aspartic acid residues and glutamic acid residues), as shown in Figure 6. Deamidation of proteins can cause protein unfolding or conformational rearrangement while reducing the formation of hydrogen bonds within or between protein subunits caused by the amide groups, which in turn reduces the aggregation between molecules or subunits and enhances the hydrogen bond or hydrophilic interaction between proteins and water molecules [49]. Additionally, deamidation increases the electrostatic repulsion between protein chains and decreases the surface hydrophobicity of the subunits [50].

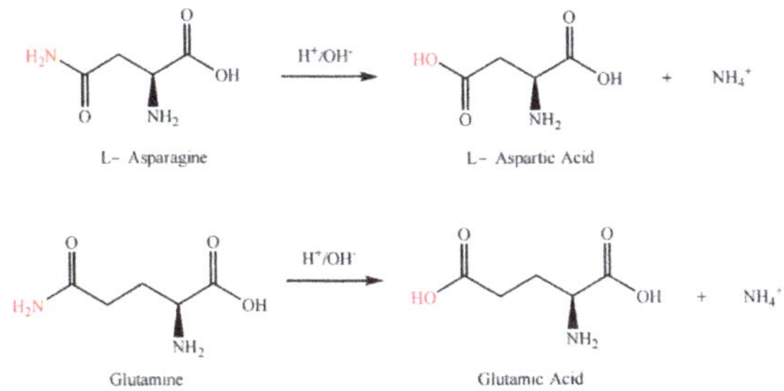

Figure 6. Deamidation reaction formula of rice protein.

Guan et al. [51] used the deamidation method under alkaline conditions to obtain highly soluble rice bran protein, by conducting a 30 min deamidation reaction at pH 8.0 and 100 °C. Another approach for deamidation modification is acid-catalyzed deamidation, which is often characterized by the mild properties of organic acids and can prevent issues such as the breakdown of peptide bonds and the isomerization of amino acid residues brought on by excessive deamidation [52]. Li et al. [53] adjusted the pH of glutelin solutions to 4.0 using malic acid and citric acid, and then subjected them to deamidation treatment, resulting in an increase in solubility of glutelin from 7.79% to 39.13% and 26.06%, respectively.

In addition, enzyme-assisted protein deamidation has high reaction specificity and food safety. This method can increase the surface electrostatic charge of the protein, reduce the hydrogen bonds within the molecule, form hydrophilic carboxyl groups, and promote protein unfolding [54]. Liu et al. [49] found that deamidation of rice glutelin by glutamine transamidase could increase its solubility in neutral or slightly acidic solutions (i.e., pH 5–7). Chen et al. [55] discovered that glutamine transamidase could prevent excessive hydrolysis caused by chemical or other protease treatments while increasing the water solubility of glutelin.

3.6. Modification and Solubility Enhancement of Rice Protein by Enzymatic Hydrolysis Method

The enzymatic hydrolysis method can break down protein into short-chain peptides or amino acids, and reduce intra- or inter molecular cross-linking, particularly the formation of inter-subunit disulfide bonds. The small molecules generated by hydrolysis not only exhibit good water solubility, but also possess biological activities such as anti-cancer, anti-hypertension, and immune regulation [56]. Table 1 shows commonly used enzymes, reaction conditions, and modification effects for modified rice protein. Among these proteases, alkaline protease has the highest degree of hydrolysis (DH), and the modification and solubilization of rice protein are the most significant, with solubility as high as 94%. This is due to the fact that alkaline protease has more enzyme hydrolysis sites, resulting in a large number of hydrolysis of hydrophobic amino acid residues, thus improving the solubility of rice protein [57].

The advantages of enzyme hydrolysis include mild reaction conditions, easy control, and high specificity. However, traditional enzymatic hydrolysis can result in bitter peptides (mainly composed of peptides containing Lys, Leu, and Val residues), reduce the emulsifying ability of protein, and has a low hydrolysis efficiency and extended reaction time [58,59]. Physical methods such as ultrasound, high pressure, or the construction of a dual-enzyme system can be used to not only speed up the enzymatic reaction time and increase the efficiency of enzymatic hydrolysis but also to reduce bitter peptides and increase the emulsification and emulsifying stability of the protein.

Yang et al. [60] show that ultrasound-assisted enzymatic hydrolysis can significantly improve the efficiency of rice protein processing. The cavitation effect generated by ultrasound can promote the stretching of enzymatic protein molecules, reduce disulfide bond content and the hydrophobicity of the protein surface, increase the dissolution rate of soluble protein particles or protein molecules, and enhance the contact frequency between protein and enzyme [61], thus improving the enzymatic hydrolysis efficiency. Chang et al. [62] found that ultrasound-assisted enzymatic hydrolysis can promote the exposure of cleavage sites for papain, reduce the hydrolysis time for 1 h, and the enzyme dosage amount by 1.4 times. At the same time, ultrasound-assisted enzymatic hydrolysis increased the water solubility of rice bran protein by nearly 2 times, while the emulsifying stability decreased by 38.25%. This is because the low molecular weight peptides formed by long-time enzymatic hydrolysis (3 h) could not stabilize the oil–water interface, which in turn has an adverse effect on protein emulsification.

High-pressure assistance can improve the solubility and emulsifying stability of rice protein by enhancing enzymatic hydrolysis. Liu et al. [63] found that a pressure of 300 MPa can increase the hydrolysis of rice protein by alkaline protease, improve the solubility of rice protein by nearly 1.7 times, and increase the emulsifying property and emulsifying stability by 2 times and 3 times, respectively. This is due to the changes in the tertiary and quaternary structure of protein molecules caused by ultra-high pressure, which releases smaller soluble proteins and promotes the modification of protease. At the same time, ultra-high pressure shortens the enzymatic hydrolysis time to 15 min and avoids the decrease in protein emulsifying ability. Zhang et al. [64] also found that high-pressure micro fluidization can promote the exposure of hydrophobic groups within the rice bran separation protein, increasing the surface hydrophobicity and molecular diffusion rate, which significantly improves the hydrolysis of rice bran separation protein by neutral protease.

The hydrophobic amino acids in rice protein participate in peptide bond formation at the -amino or carboxyl end during enzymatic hydrolysis, giving the hydrolysate an unfavorable bitter taste [65]. Yan et al. [66] constructed a dual-enzyme system of aminopeptidase and pancreatin, which reduced the bitterness of rice protein hydrolysates. Pooja et al. [67] found that the hydrolysate of rice bran protein pretreated with high hydrostatic pressure can also effectively address issues such as prolonged hydrolysis time and bitter peptides.

Table 1. Common enzymes, reaction conditions, and modification effect of modified rice protein.

Enzyme	Condition	Targets	DH	Effect	References
Papain	Enzyme: substrate = 3:100; 4 h; pH = 7.0; 50 °C	Carboxyl terminus of arginine, lysine, and glycine residues	15–32%	Increased the solubility (about 45–94%)	[68]
Trypsin	Enzyme: substrate = 0.89:1000; 2.4 h; pH = 7.6; 52.8 °C	Carboxyl terminal of arginine and lysine residues	8.96%	Increased the solubility (above 75%)	[65]
Alkaline protease	Enzyme: substrate = 1:100; 5 h; pH = 8.0; 65 °C	Carboxyl of hydrophobic amino acid–amide bond of aromatic amino acids	23.8%	Increased the solubility (to 94.78%)	[69,70]
Glutaminase	Enzyme: substrate = 1:250; 12 h; pH = 8.0; 50 °C	Acyl transfer reaction between lysine residue and glutamine residue	4–6%	Increased the solubility (to 78.14%)	[49,71]

4. Applications of Modified Rice Protein in Food Processing

Modified rice protein exhibits desirable functional properties such as high solubility, emulsifying ability, gelling ability, and antioxidant activity, which make it a promising ingredient for a wide range of food applications. Modified rice protein can replace allergenic bovine milk protein in dairy products, increasing the potential for developing plant-based dairy products or infant formula [72]. In meat products, it can balance the nutritional value, reduce economic costs, and improve product stability [73]. In baked goods, it can be used to develop gluten-free baked goods, providing more options for people with gluten allergies [74].

4.1. Application of Modified Rice Protein in Dairy Product Processing

Modified rice protein retains its low allergenicity, and its emulsifying and encapsulating properties enable the inclusion of vitamins, minerals, probiotics, etc., making it useful in developing infant formula, probiotic fermented milk, cheese, and other dairy products [75].

Modified rice protein can be used in fermented milk products, taking advantage of its good emulsifying and encapsulating properties [76]. Vaniski et al. [77] found that the encapsulation efficiency of thermophilic streptococci by rice bran protein–maltodextrin covalent complexes can reach 90.26%, and the survival rate of thermophilic streptococci in simulated gastric and intestinal fluids is relatively high. Zhang et al. [78] prepared rice protein–pectin composite microcapsules, which exhibited excellent antibacterial activity and could inhibit key enzymes in the tricarboxylic acid cycle and hexose monophosphate pathway of Escherichia coli. After encapsulating probiotics with modified rice protein, it can be directly added as an ingredient to fermented dairy products, and protect the activity of probiotics during food processing and digestion in the human gastrointestinal tract.

Rice protein has low allergenicity and is often used as a substitute for cow's milk protein in formula powders for lactose intolerant individuals, with certain digestive tolerance, safety, acceptability, and palatability [79]. Amagliani et al. [80] added low-molecular-weight surfactants to hydrolyzed rice protein, mixed it with oils, carbohydrates, and maltodextrins, and developed a rice protein infant formula emulsion formula. The formula reduced the size of fat globules and had high emulsion and thermal stability.

Modified rice protein can also be applied in cheese products, mainly as an active filler embedded in the cheese protein matrix to produce high-quality low-fat cheese [81]. Paximada et al. [82] prepared a water-in-oil (W/O) emulsion of modified rice protein with fat and then homogenized the emulsion with milk to form a water-in-oil-in-water (W/O/W) double emulsion. The process flow diagram of preparing cheese using the W/O/W double emulsion is shown in Figure 7. The double emulsion has a high protein encapsulation efficiency, which can reduce the loss of fat in cheese products and decrease the hardness, thus having a broad application prospect for developing low-fat dairy products. In addition to forming stable water-in-oil emulsions with lipids, the improved functional properties of modified rice protein also facilitate protein gel formation during fermentation, providing possibilities for the development of fermented plant-based cheese [72].

Figure 7. The flow diagram of W/O/W double emulsion preparation of cheese.

4.2. Application of Modified Rice Protein in Meat Products Processing

The processing of meat products is often accompanied by the loss of intracellular and extracellular juice in muscle tissue, as well as a decrease in the water-holding capacity of muscle proteins. Excessive intake of animal protein can also lead to cardiovascular diseases [83] and kidney problems [84]. Modified rice protein has good water-holding properties, which can reduce the loss of fluid in muscle tissue and balance the nutritional value of animal protein, thereby reducing the negative effects of consuming excessive saturated fatty acids on the body [85]. At present, modified rice protein is commonly used in block meat, minced meat, and plant-based meat products.

In the processing of block meat products, modified rice protein can reduce the dehydration shrinkage of whole or large pieces of meat, maintain the integrity of the meat muscle tissue, and improve the product yield. Aqsa et al. [86] mixed chicken chunks with modified rice protein isolate by kneading. The results showed that modified rice protein isolate can increase the protein content of chicken chunks (from 34.99% to 48.49%) and reduce cooking loss (from 12.44% to 3.85%). Zhou et al. [69] found that modified rice protein has a strong antioxidant ability and can effectively reduce lipid oxidation in meat during storage.

Modified rice protein can reduce the loss of fat and moisture in minced meat products and improve sensory quality [87]. Li et al. [88] found that alkaline protease-modified rice protein can form a dense covering film on the surface of the muscle fibers of the sausage. Meanwhile, modified rice protein can form an elastic and hard gel network with meat protein, which can lock fat and moisture in the three-dimensional network structure, reduce the loss of juice, and increase the yield of sausage [89].

Modified rice protein can also be used in the production of plant-based meat products. It has low allergenicity, no beany flavor, and a fatty mouthfeel, and can partially replace soy protein in the production of plant-based meat products [90]. Lee et al. [91] mixed modified rice protein and soy protein in a certain proportion, and used corn starch and wheat flour as additives. They employed a low-moisture extrusion-cooking process with a twin-screw extruder to promote the interaction between proteins, lipids, and carbohydrates, resulting in the production of plant-based meat products with high nutritional value and unique flavor.

4.3. Application of Modified Rice Protein in Baked Food Processing

Baked food is typically made from wheat flour through processes such as kneading, fermentation, and baking. However, wheat contains allergenic components such as gluten, which makes it difficult for patients with celiac disease or wheat gluten protein allergies to consume. Therefore, there is a need to develop gluten-free baked food to meet the needs of these special populations. The modified rice protein is free of gluten, has low allergenicity, and has a high water absorption and oil-holding capacity. It can successfully lower the loss of water and oil in baked goods, keep the texture fresh and moist, and raise the amount of protein in them [74].

When mixed with water, sugar, oil, salt, and other ingredients in a certain proportion, modified rice protein can form a sticky and elastic network similar to gluten, which can be baked at high temperatures to make bread, biscuits, or cakes. Sahagún et al. [92] used a recipe containing rice flour, modified rice protein, sugar, milk, pasteurized egg liquid, sunflower oil, and baking powder to make cakes, which improved the texture characteristics of the cakes. Yadav et al. [93] used 5% rice bran concentrate protein instead of refined wheat flour to prepare biscuits, and the results showed that as the concentration of rice bran concentrate protein increased, the physical characteristics of the biscuits, such as diameter, thickness, fracture strength, moisture content, protein content, and ash content, all increased significantly. This may be because rice bran concentrate protein absorbed more water and oil during the baking process, maintaining the moist texture of the biscuits and giving them the best texture in terms of color, taste, and flavor. Honda et al. [94] found that after protease modification, rice protein and gluten protein formed a three-dimensional structure, which increased the volume of gluten-free bread by 22%. Scanning electron microscopy revealed that both the size and quantity of pores in the bread increased, leading to improvements in its rheological and sensory properties [95].

5. Conclusions

The modification and solubilization of rice protein are very significant; however, the traditional modification of rice protein is often accompanied by some issues, such as long reaction time, plenty of by-products, insufficient reaction, and so on, while the improved modification method avoids these shortcomings. Modification and solubilization of rice protein by polysaccharides, polyphenols, heterologous proteins, and phosphates are assisted by ultrasound, ultra-high pressure, microwave, enzymes, water bath heating, or alkaline conditions. The modification and solubilization of rice protein by deamidation are assisted by mild organic acids or enzyme-assisted deamidation. However, the control of the degree of deamidation and the safety of protein are still the difficulties of current research. Enzymatic hydrolysis for the modification and solubilization of rice protein is assisted by ultrasound, ultra-high pressure, and dual-enzyme systems. However, it is difficult to realize industrial applications due to the high cost of the experiment. To summarize, under certain physical and chemical conditions, combining rice protein with a certain amount of polysaccharides, polyphenols, heterologous protein, or phosphate anions is the most simple, safe, effective, and low-cost modification method. Among these methods, the polysaccharide reaction can be carried out spontaneously without other chemical reagents, so it has more application potential in improving the water solubility of rice protein.

Modified rice protein has been widely used in the processing of dairy products, meat products, and baked goods due to its good solubility, emulsifying properties, gelling properties, water-holding capacity, and oil-holding capacity. It not only balances nutritional value and meets the protein intake needs of allergic populations, but also improves the deficiencies of animal and plant proteins, increases stability, and extends shelf life. With the in-depth research on the functional properties of modified rice protein, its application scope in food processing is constantly expanding, which can increasingly meet the needs of the food industry and market sectors.

Author Contributions: J.Y.: Collected the literature sources and Wrote the manuscript. D.M.: Conceptualized the idea and Performed critical reviews on the manuscript. Z.W.: Formal analysis, Supervision. Z.W., J.C. and L.X.: Review & Editing, Funding acquisition. All authors have read and agreed to the published version of the manuscript.

Funding: This research was supported by the Natural Science Foundation of Tianjin (No. 22JCQNJC01320), National Natural Science Foundation of China (No. 31701612), National Training Programs for Innovation and Entrepreneurship of Undergraduates of China (202110069018 and 202210069004), Tianjin "131" innovative talent team (Team number 201927), Tianjin Key Research and Development Program and Tianjin Applied Basic and Frontier Technology Research Program (19YFLHSN00080, S19FC2020 and 19JCTPJC54600), the Key Research and Development Plan of Anhui Province (202204c06020013), and the Anhui Provincial Excellent Research and Innovation Team Project.

Institutional Review Board Statement: Not applicable.

Informed Consent Statement: Not applicable.

Data Availability Statement: Data are contained within the article.

Conflicts of Interest: The authors declare no conflict of interest.

Sample Availability: Not applicable.

References

1. Wen, L.; Huang, L.; Li, Y.; Feng, Y.; Zhang, Z.; Xu, Z.; Chen, M.-L.; Cheng, Y. New peptides with immunomodulatory activity identified from rice proteins through peptidomic and in silico analysis. *Food Chem.* **2021**, *364*, 130357. [CrossRef]
2. Lin, D.; Zhang, L.; Li, R.; Zheng, B.; Rea, M.C.; Miao, S. Effect of plant protein mixtures on the microstructure and rheological properties of myofibrillar protein gel derived from red sea bream (Pagrosomus major). *Food Hydrocoll.* **2019**, *96*, 537–545. [CrossRef]
3. Piotrowicz, I.B.B.; Garcés-Rimón, M.; Moreno-Fernández, S.; Aleixandre, A.; Salas-Mellado, M.; Miguel-Castro, M. Antioxidant, Angiotensin-Converting Enzyme Inhibitory Properties and Blood-Pressure-Lowering Effect of Rice Bran Protein Hydrolysates. *Foods* **2020**, *9*, 812. [CrossRef] [PubMed]
4. Ren, L.; Fan, J.; Yang, Y.; Xu, Y.; Chen, F.; Bian, X.; Xing, T.; Liu, L.; Yu, D.; Zhang, N. Enzymatic Hydrolysis of Broken Rice Protein: Antioxidant Activities by Chemical and Cellular Antioxidant Methods. *Front. Nutr.* **2021**, *8*, 788078. [CrossRef]
5. Yu, S.; Fang, N.; Li, Q.; Zhang, J.; Luo, H.; Ronis, M.; Badger, T.M. In vitro actions on human cancer cells and the liquid chromatography-mass spectrometry/mass spectrometry fingerprint of phytochemicals in rice protein isolate. *J. Agric. Food Chem.* **2006**, *54*, 4482–4492. [CrossRef]
6. Zhang, H.; Bartley, G.E.; Mitchell, C.R.; Zhang, H.; Yokoyama, W. Lower weight gain and hepatic lipid content in hamsters fed high fat diets supplemented with white rice protein, brown rice protein, soy protein, and their hydrolysates. *J. Agric. Food Chem.* **2011**, *59*, 10927. [CrossRef]

7. Yousefi, N.; Abbasi, S. Food proteins: Solubility & thermal stability improvement techniques. *Food Chem. Adv.* **2022**, *1*, 100090. [CrossRef]
8. Zhao, M.; Xiong, W.; Chen, B.; Zhu, J.; Wang, L. Enhancing the solubility and foam ability of rice glutelin by heat treatment at pH12: Insight into protein structure. *Food Hydrocoll.* **2020**, *103*, 105626. [CrossRef]
9. Zhong, L.; Ma, N.; Wu, Y.; Zhao, L.; Ma, G.; Pei, F.; Hu, Q. Characterization and functional evaluation of oat protein isolate-*Pleurotus ostreatus* β-glucan conjugates formed via Maillard reaction. *Food Hydrocoll.* **2019**, *87*, 459–469. [CrossRef]
10. Thongkong, S.; Klangpetch, W.; Unban, K.; Tangjaidee, P.; Phimolsiripol, Y.; Rachtanapun, P.; Jantanasakulwong, K.; Schönlechner, R.; Thipchai, P.; Phongthai, S. Impacts of Electroextraction Using the Pulsed Electric Field on Properties of Rice Bran Protein. *Foods* **2023**, *12*, 835. [CrossRef]
11. Christian, X.W.; Manfred, J.S. Self-Consistent Assignment of Asparagine and Glutamine Amide Rotamers in Protein Crystal Structures. *Structure* **2006**, *14*, 967–972. [CrossRef]
12. Li, T.; Wang, L.; Chen, Z.; Sun, D.; Li, Y. Electron beam irradiation induced aggregation behaviour, structural and functional properties changes of rice proteins and hydrolysates. *Food Hydrocoll.* **2019**, *97*, 105192. [CrossRef]
13. Robert, P.B. A tensegrity model for hydrogen bond networks in proteins. *Heliyon* **2017**, *3*, e307. [CrossRef]
14. He, R.; He, H.-Y.; Chao, D.; Ju, X.; Aluko, R. Effects of High Pressure and Heat Treatments on Physicochemical and Gelation Properties of Rapeseed Protein Isolate. *Food Bioprocess Technol.* **2014**, *7*, 1344–1353. [CrossRef]
15. Liu, P.; Li, Y.; Gao, L.; Zhou, X.; Ma, P.; Wang, Q. Effect of different carbohydrates on the functional properties of black rice glutelin (BRG) modified by the maillard reaction. *J. Cereal Sci.* **2020**, *93*, 102979. [CrossRef]
16. Perusko, M.; Al-Hanish, A.; Velickovic, T.C.; Stanic-Vucinic, D. Macromolecular crowding conditions enhance glycation and oxidation of whey proteins in ultrasound-induced Maillard reaction. *Food Chem.* **2015**, *177*, 248–257. [CrossRef]
17. Li, M.; McClements, D.J.; Liu, X.; Liu, F. Design principles of oil-in-water emulsions with functionalized interfaces: Mixed, multilayer, and covalent complex structures. *Compr. Rev. Food Sci. Food Saf.* **2020**, *19*, 3159–3190. [CrossRef]
18. Kan, X.; Chen, G.; Zhou, W.; Zeng, X. Application of protein-polysaccharide Maillard conjugates as emulsifiers: Source, preparation and functional properties. *Food Res. Int.* **2021**, *150*, 110740. [CrossRef]
19. Zhifeng, Z.; Youling, L.X. Thermosonication-induced structural changes and solution properties of mung bean protein. *Ultrason. Sonochem.* **2019**, *62*, 104908. [CrossRef]
20. Chen, X.; Zhao, H.; Wang, H.; Xu, P.; Chen, M.; Xu, Z.; Wen, L.; Cui, B.; Yu, B.; Zhao, H.; et al. Preparation of high-solubility rice protein using an ultrasound-assisted glycation reaction. *Food Res. Int.* **2022**, *161*, 111737. [CrossRef]
21. Liu, X.; Yang, Q.; Yang, M.; Du, Z.; Wei, C.; Zhang, T.; Liu, B.; Liu, J. Ultrasound-assisted Maillard reaction of Ovalbumin/Xylose: The enhancement of functional properties and its mechanism. *Ultrason. Sonochem.* **2021**, *73*, 105477. [CrossRef]
22. Jia, X.; Li, L.; Teng, J.; Li, M.; Long, H.; Xia, N. Glycation of rice protein and d-xylose pretreated through hydrothermal cooking-assisted high hydrostatic pressure: Focus on the structural and functional properties. *LWT* **2022**, *160*, 113194. [CrossRef]
23. Cheng, Y.-H.; Mu, D.-C.; Jiao, Y.; Xu, Z.; Chen, M.-L. Microwave-assisted maillard reaction between rice protein and dextran induces structural changes and functional improvements. *J. Cereal Sci.* **2021**, *97*, 103134. [CrossRef]
24. Meng, X.; Li, T.; Song, T.; Chen, C.; Venkitasamy, C.; Pan, Z.; Zhang, H. Solubility, structural properties, and immunomodulatory activities of rice dreg protein modified with sodium alginate under microwave heating. *Food Sci. Nutr.* **2019**, *7*, 2556–2564. [CrossRef]
25. Zheng, Y.; Li, Z.; Lu, Z.; Wu, F.; Fu, G.; Zheng, B.; Tian, Y. Structural characteristics and emulsifying properties of lotus seed protein isolate-dextran glycoconjugates induced by a dynamic high pressure microfluidization Maillard reaction. *LWT* **2022**, *160*, 113309. [CrossRef]
26. Li, Y.; He, D.; Li, B.; Lund, M.N.; Xing, Y.; Wang, Y.; Li, F.; Cao, X.; Liu, Y.; Chen, X.; et al. Engineering polyphenols with biological functions via polyphenol-protein interactions as additives for functional foods. *Trends Food Sci. Technol.* **2021**, *110*, 470–482. [CrossRef]
27. Keppler, J.K.; Schwarz, K.; van der Goot, A.J. Covalent modification of food proteins by plant-based ingredients (polyphenols and organosulphur compounds): A commonplace reaction with novel utilization potential. *Trends Food Sci. Technol.* **2020**, *101*, 38–49. [CrossRef]
28. Dai, T.; Yan, X.; Li, Q.; Li, T.; Liu, C.; McClements, D.J.; Chen, J. Characterization of binding interaction between rice glutelin and gallic acid: Multi-spectroscopic analyses and computational docking simulation. *Food Res. Int.* **2017**, *102*, 274–281. [CrossRef]
29. Wang, S.; Li, X.; Zhu, J.; Liu, H.; Liu, T.; Yu, G.; Shao, M. Covalent Interaction between High Hydrostatic Pressure-Pretreated Rice Bran Protein Hydrolysates and Ferulic Acid: Focus on Antioxidant Activities and Emulsifying Properties. *J. Agric. Food Chem.* **2021**, *69*, 7777–7785. [CrossRef]
30. Quan, T.H.; Benjakul, S.; Sae-Leaw, T.; Balange, A.K.; Maqsood, S. Protein–polyphenol conjugates: Antioxidant property, functionalities and their applications. *Trends Food Sci. Technol.* **2019**, *91*, 507–517. [CrossRef]
31. Jaroslaw, C.; Krzysztof, D. A review of methods used for investigation of protein–phenolic compound interactions. *Int. J. Food Sci. Technol.* **2017**, *52*, 573–585. [CrossRef]
32. Shi-Qi, T.; Qiu-Han, D.; Zhen, F. Ultrasonic Treatment on Physicochemical Properties of Water-Soluble Protein from Moringa Oleifera Seed. *Ultrason. Sonochem.* **2020**, *71*, 105357. [CrossRef]
33. Xue, F.; Xie, Y.; Li, C.; Wang, S.; Liu, X. Prevention of frozen-dough from deterioration with incorporation of glutenin-polyphenols conjugates prepared by ultrasound. *LWT* **2021**, *151*, 112141. [CrossRef]

34. Wang, T.; Yue, M.; Xu, P.; Wang, R.; Chen, Z. Toward water-solvation of rice proteins via backbone hybridization by casein. *Food Chem.* **2018**, *258*, 278–283. [CrossRef] [PubMed]
35. Wang, R.; Li, L.; Feng, W.; Wang, T. Fabrication of hydrophilic composites by bridging the secondary structures between rice proteins and pea proteins toward enhanced nutritional properties. *Food Funct.* **2020**, *11*, 7446–7455. [CrossRef]
36. Wang, R.; Xu, P.; Chen, Z.; Zhou, X.; Wang, T. Complexation of rice proteins and whey protein isolates by structural interactions to prepare soluble protein composites. *LWT* **2019**, *101*, 207–213. [CrossRef]
37. Li, F.; Wang, T.; Feng, W.; Wang, R.; Chen, Z.; Yi, D. Novel Protein Hydrocolloids Constructed by Hydrophobic Rice Proteins and Walnut Proteins as Loading Platforms for Nutraceutical Models. *Food Biophys.* **2021**, *16*, 427–439. [CrossRef]
38. Baek, M.; Mun, S. Improvement of the water solubility and emulsifying capacity of rice proteins through the addition of isolated soy protein. *Int. J. Food Sci. Technol.* **2022**, *57*, 4411–4421. [CrossRef]
39. Kuddus, M. Transglutaminase protein substrates of food interest. In *Enzymes in Food Technology*; Springer: Singapore, 2018; pp. 293–317. [CrossRef]
40. He, C.; Hu, Y.; Woo, M.W.; Xiong, H.; Zhao, Q. Effect of microbial transglutaminase on the structural and rheological characteristics and in vitro digestion of rice glutelin–casein blends. *Food Res. Int.* **2020**, *139*, 109832. [CrossRef]
41. Pei, Y.; Wan, J.; You, M.; McClements, D.J.; Li, Y.; Li, B. Impact of whey protein complexation with phytic acid on its emulsification and stabilization properties. *Food Hydrocoll.* **2019**, *87*, 90–96. [CrossRef]
42. Zhouyi, X.; Maojie, Z.; Meihu, M. Emulsifying properties of ovalbumin: Improvement and mechanism by phosphorylation in the presence of sodium tripolyphosphate. *Food Hydrocoll.* **2016**, *60*, 29–37. [CrossRef]
43. Ma, Z.; Chi, Y.; Zhang, H.; Chi, Y.; Ma, Y. Inhibiting effect of dry heat on the heat-induced aggregation of egg white protein. *Food Chem.* **2022**, *387*, 132850. [CrossRef]
44. Li, P.; Sun, Z.; Ma, M.; Jin, Y.; Sheng, L. Effect of microwave-assisted phosphorylation modification on the structural and foaming properties of egg white powder. *LWT* **2018**, *97*, 151–156. [CrossRef]
45. Hu, Z.; Qiu, L.; Sun, Y.; Xiong, H.; Ogra, Y. Improvement of the solubility and emulsifying properties of rice bran protein by phosphorylation with sodium trimetaphosphate. *Food Hydrocoll.* **2019**, *96*, 288–299. [CrossRef]
46. Peishan, L.; Yongguo, J.; Long, S. Impact of microwave assisted phosphorylation on the physicochemistry and rehydration behaviour of egg white powder. *Food Hydrocoll.* **2020**, *100*, 105380. [CrossRef]
47. Hu, Y.; Du, L.; Sun, Y.; Zhou, C.; Pan, D. Recent developments in phosphorylation modification on food proteins: Structure characterization, site identification and function. *Food Hydrocoll.* **2023**, *137*, 108390. [CrossRef]
48. Hadidi, M.; Jafarzadeh, S.; Ibarz, A. Modified mung bean protein: Optimization of microwave-assisted phosphorylation and its functional and structural characterizations. *LWT* **2021**, *151*, 112119. [CrossRef]
49. Liu, Y.; Li, X.; Zhou, X.; Yu, J.; Wang, F.; Wang, J. Effects of glutaminase deamidation on the structure and solubility of rice glutelin. *LWT-Food Sci. Technol.* **2011**, *44*, 2205–2210. [CrossRef]
50. Fang, L.; Xiang, H.; Sun-Waterhouse, D.; Cui, C.; Lin, J. Enhancing the Usability of Pea Protein Isolate in Food Applications through Modifying Its Structural and Sensory Properties via Deamidation by Glutaminase. *J. Agric. Food Chem.* **2020**, *68*, 1691–1697. [CrossRef]
51. Guan, J.; Takai, R.; Toraya, K.; Ogawa, T.; Muramoto, K.; Mohri, S.; Ishikawa, D.; Fujii, T.; Chi, H.; Cho, S.-J. Effects of Alkaline Deamidation on the Chemical Properties of Rice Bran Protein. *Food Sci. Technol. Res.* **2017**, *23*, 697–704. [CrossRef]
52. Liao, L.; Liu, T.-X.; Zhao, M.-M.; Cui, C.; Yuan, B.-E.; Tang, S.; Yang, F. Functional, nutritional and conformational changes from deamidation of wheat gluten with succinic acid and citric acid. *Food Chem.* **2010**, *123*, 123–130. [CrossRef]
53. Lei, L.; Zhao, Q.; Selomulya, C.; Xiong, H. The effect of deamidation on the structural, functional, and rheological properties of glutelin prepared from Akebia trifoliata var. australis seed. *Food Chem.* **2015**, *178*, 96–105. [CrossRef]
54. Liu, X.; Wang, C.; Zhang, X.; Zhang, G.; Zhou, J.; Chen, J. Application Prospect of Protein-Glutaminase in the Development of Plant-Based Protein Foods. *Foods* **2022**, *11*, 440. [CrossRef] [PubMed]
55. Chen, X.; Fu, W.; Luo, Y.; Cui, C.; Suppavorasatit, I.; Liang, L. Protein deamidation to produce processable ingredients and engineered colloids for emerging food applications. *Compr. Rev. Food Sci. Food Saf.* **2021**, *20*, 3788–3817. [CrossRef]
56. García, M.; Puchalska, P.; Esteve, C.; Marina, M. Vegetable foods: A cheap source of proteins and peptides with antihypertensive, antioxidant, and other less occurrence bioactivities. *Talanta* **2013**, *106*, 328–349. [CrossRef] [PubMed]
57. Doucet, D.; Otter, D.E.; Gauthier, S.F.; Foegeding, E.A. Enzyme-induced gelation of extensively hydrolyzed whey proteins by Alcalase: Peptide identification and determination of enzyme specificity. *J. Agric. Food Chem.* **2003**, *51*, 6300–6308. [CrossRef]
58. Meinlschmidt, P.; Sussmann, D.; Schweiggert-Weisz, U.; Eisner, P. Enzymatic treatment of soy protein isolates: Effects on the potential allergenicity, technofunctionality, and sensory properties. *Food Sci. Nutr.* **2016**, *4*, 11–23. [CrossRef]
59. Zang, X.; Yue, C.; Wang, Y.; Shao, M.; Yu, G. Effect of limited enzymatic hydrolysis on the structure and emulsifying properties of rice bran protein. *J. Cereal Sci.* **2019**, *85*, 168–174. [CrossRef]
60. Yang, X.; Wang, L.; Zhang, F.; Ma, H. Effects of multi-mode S-type ultrasound pretreatment on the preparation of ACE inhibitory peptide from rice protein. *Food Chem.* **2020**, *331*, 127216. [CrossRef]
61. Zhang, Z.; Wang, Y.; Jiang, H.; Dai, C.; Xing, Z.; Mintah, B.K.; Dabbour, M.; He, R.; Ma, H. Effect of dual-frequency ultrasound on the formation of lysinoalanine and structural characterization of rice dreg protein isolates. *Ultrason. Sonochem.* **2020**, *67*, 105124. [CrossRef]

62. Chang, H. *Improving Emulsibility Properties of Rice Bran Protein with Ultrasonic and Papain Hydrolysis Methods*; Henan University of Technology: Zhengzhou, China, 2019. [CrossRef]
63. Liu, N.; Lin, P.; Zhang, K.; Yao, X.; Li, D.; Yang, L.; Zhao, M. Combined effects of limited enzymatic hydrolysis and high hydrostatic pressure on the structural and emulsifying properties of rice proteins. *Innov. Food Sci. Emerg. Technol.* **2022**, *77*, 102975. [CrossRef]
64. Zhang, L.; Chen, X.; Wang, Y.; Guo, F.; Hu, S.; Hu, J.; Xiong, H.; Zhao, Q. Characteristics of rice dreg protein isolate treated by high-pressure microfluidization with and without proteolysis. *Food Chem.* **2021**, *358*, 129861. [CrossRef] [PubMed]
65. Li, X.; Xiong, H.; Yang, K.; Peng, D.; Peng, H.; Zhao, Q. Optimization of the biological processing of rice dregs into nutritional peptides with the aid of trypsin. *J. Food Sci. Technol.-Mysore* **2012**, *49*, 537–546. [CrossRef]
66. Yan, Z.-F.; Yuan, S.; Qin, Q.; Wu, J. Enhancement of rice protein hydrolysate quality using a novel dual enzyme system. *LWT* **2022**, *158*, 113110. [CrossRef]
67. Pooja, K.; Rani, S.; Kanwate, B.; Pal, G.K. Physico-chemical, Sensory and Toxicity Characteristics of Dipeptidyl Peptidase-IV Inhibitory Peptides from Rice Bran-derived Globulin Using Computational Approaches. *Int. J. Pept. Res. Ther.* **2017**, *23*, 519–529. [CrossRef]
68. Singh, T.P.; Siddiqi, R.A.; Sogi, D.S. Enzymatic modification of rice bran protein: Impact on structural, antioxidant and functional properties. *LWT-Food Sci. Technol.* **2021**, *138*, 110648. [CrossRef]
69. Zhou, K.; Canning, C.; Sun, S. Effects of rice protein hydrolysates prepared by microbial proteases and ultrafiltration on free radicals and meat lipid oxidation. *LWT-Food Sci. Technol.* **2013**, *50*, 331–335. [CrossRef]
70. Braspaiboon, S.; Osiriphun, S.; Peepathum, P.; Jirarattanarangsri, W. Comparison of the effectiveness of alkaline and enzymatic extraction and the solubility of proteins extracted from carbohydrate-digested rice. *Heliyon* **2020**, *6*, e5403. [CrossRef]
71. Hu, Y.; Sun-Waterhouse, D.; Liu, P.; Cui, C.; Wang, W. Modification of rice protein with glutaminase for improved structural and sensory properties. *Int. J. Food Sci. Technol.* **2019**, *54*, 2458–2467. [CrossRef]
72. Masiá, C.; Jensen, P.E.; Petersen, I.L.; Buldo, P. Design of a Functional Pea Protein Matrix for Fermented Plant-Based Cheese. *Foods* **2022**, *11*, 178. [CrossRef]
73. Owusu-Ansah, P.; Besiwah, E.K.; Bonah, E.; Amagloh, F.K. Non-meat ingredients in meat products: A scoping review. *Appl. Food Res.* **2022**, *2*, 100044. [CrossRef]
74. Chandi, G.K.; Sogi, D.S. Functional properties of rice bran protein concentrates. *J. Food Eng.* **2007**, *79*, 592–597. [CrossRef]
75. Nunes, L.; Martins, E.; Perrone, T.; De Carvalho, A.F. The Maillard Reaction in Powdered Infant Formula. *J. Food Nutr. Res.* **2019**, *7*, 33–40. [CrossRef]
76. Gruskiene, R.; Bockuviene, A.; Sereikaite, J. Microencapsulation of Bioactive Ingredients for Their Delivery into Fermented Milk Products: A Review. *Molecules* **2021**, *26*, 4601. [CrossRef] [PubMed]
77. Vaniski, R.; da Silva, S.C.; Silva-Buzanello, R.A.; Canan, C.; Drunkler, D.A. Improvement of Lactobacillus acidophilus La-5 microencapsulation viability by spray-drying with rice bran protein and maltodextrin. *J. Food Process. Preserv.* **2021**, *45*, 15364. [CrossRef]
78. Zhang, L.; Zhang, M.; Ju, R.; Bhandari, B.; Liu, K. Antibacterial mechanisms of star anise essential oil microcapsules encapsulated by rice protein-depolymerized pectin electrostatic complexation and its application in crab meatballs. *Int. J. Food Microbiol.* **2023**, *384*, 109963. [CrossRef]
79. Bocquet, A.; Dupont, C.; Chouraqui, J.-P.; Darmaun, D.; Feillet, F.; Frelut, M.-L.; Girardet, J.-P.; Hankard, R.; Lapillonne, A.; Rozé, J.-C.; et al. Efficacy and safety of hydrolyzed rice-protein formulas for the treatment of cow's milk protein allergy. *Arch. De Pediatr.* **2019**, *26*, 238–246. [CrossRef]
80. Amagliani, L.; O'Regan, J.; Kelly, A.L.; O'Mahony, J.A. Influence of low molecular weight surfactants on the stability of model infant formula emulsions based on hydrolyzed rice protein. *LWT-Food Sci. Technol.* **2022**, *154*, 112544. [CrossRef]
81. Ramel, P.R.; Marangoni, A.G. Processed cheese as a polymer matrix composite: A particle toolkit for the replacement of milk fat with canola oil in processed cheese. *Food Res. Int.* **2018**, *107*, 110–118. [CrossRef] [PubMed]
82. Paximada, P.; Howarth, M.; Dubey, B.N. Double emulsions fortified with plant and milk proteins as fat replacers in cheese. *J. Food Eng.* **2021**, *288*, 110229. [CrossRef]
83. Qi, X.X.; Shen, P. Associations of dietary protein intake with all-cause, cardiovascular disease, and cancer mortality: A systematic review and meta-analysis of cohort studies. *Nutr. Metab. Cardiovasc. Dis.* **2020**, *30*, 1094–1105. [CrossRef] [PubMed]
84. Kramer, H. Diet and Chronic Kidney Disease. *Adv. Nutr.* **2019**, *10* (Suppl. 4), S367–S379. [CrossRef] [PubMed]
85. Sim, S.Y.J.; Srv, A.; Chiang, J.H.; Henry, C.J. Plant Proteins for Future Foods: A Roadmap. *Foods* **2021**, *10*, 1967. [CrossRef]
86. Shoaib, A.; Sahar, A.; Sameen, A.; Saleem, A.; Tahir, A.T. Use of pea and rice protein isolates as source of meat extenders in the development of chicken nuggets. *J. Food Process. Preserv.* **2018**, *42*, e13763. [CrossRef]
87. Öztürk-Kerimoğlu, B. A promising strategy for designing reduced-fat model meat emulsions by utilization of pea protein-agar agar gel complex. *Food Struct.* **2021**, *29*, 100205. [CrossRef]
88. Li, Q.; Wang, X. Study on functional properties of rice protein hydrolyzed by alkaline protease. *Guangxi Light Ind.* **2006**, *6*, 17–18. [CrossRef]
89. Yang, Q.; Tang, S.; Wei, J. Effects of different calcium salt, plant protein and dietary fiber on texture properties of ground yak meat. *Food Sci. Technol.* **2018**, *43*, 142–148. [CrossRef]

90. Sasimaporn, S.; Gi, H.R. Physicochemical and functional characteristics of plant protein-based meat analogs. *J. Food Process. Preserv.* **2019**, *43*, e14123. [CrossRef]
91. Lee, J.-S.; Oh, H.; Choi, I.; Yoon, C.S.; Han, J. Physico-chemical characteristics of rice protein-based novel textured vegetable proteins as meat analogues produced by low-moisture extrusion cooking technology. *LWT* **2022**, *157*, 113056. [CrossRef]
92. Sahagún, M.; Bravo-Núñez, Á.; Báscones, G.; Gómez, M. Influence of protein source on the characteristics of gluten-free layer cakes. *LWT-Food Sci. Technol.* **2018**, *94*, 50–56. [CrossRef]
93. Yadav, R.B.; Yadav, B.S.; Chaudhary, D. Extraction, characterization and utilization of rice bran protein concentrate for biscuit making. *Br. Food J.* **2011**, *113*, 1173–1182. [CrossRef]
94. Honda, Y.; Inoue, N.; Kurita, M.; Okunishi, T. Alpha-glutelin degradation and its hydrolysate by protease enhance the specific volume of gluten-free rice starch bread. *J. Cereal Sci.* **2021**, *102*, 103338. [CrossRef]
95. Pico, J.; Reguilón, M.P.; Bernal, J.; Gómez, M. Effect of rice, pea, egg white and whey proteins on crust quality of rice flour-corn starch based gluten-free breads. *J. Cereal Sci.* **2019**, *86*, 92–101. [CrossRef]

molecules

Article

Impact of Ultrasonication on the Self-Assembly Behavior and Gel Properties of Bovine Bone Collagen I

Hong Liu [1], Hongru Zhang [1,2], Kangyu Wang [1], Liwei Qi [1], Yujie Guo [1,*], Chunhui Zhang [1,*] and Yang Xu [3]

1 Comprehensive Key Laboratory of Agro-Products Processing, Ministry of Agriculture and Rural Affairs, Institute of Food Science and Technology, Chinese Academy of Agricultural Sciences, Beijing 100193, China
2 Laboratory of Biomass and Green Technologies, University of Liege-Gembloux Agro-Bio Tech, Passage des Déportés 2, B-5030 Gembloux, Belgium
3 Inner Mongolia Mengtai Biological Engineering Co., Ltd., Hohhot 010000, China
* Correspondence: guoyujie@caas.cn (Y.G.); zhangchunhui@caas.cn (C.Z.); Tel./Fax: +86-10-62819430 (Y.G.)

Abstract: This study deliberated the effect of ultrasonic treatment on collagen self-assembly behavior and collagen fibril gel properties. Bovine bone collagen I which had undergone ultrasonic treatment with different power (0–400 W) and duration (0–60 min) was analyzed. SDS-PAGE and spectroscopic analysis revealed that ultrasonic treatment decreased collagen molecular order degree and the number of hydrogen bonds, stretching collagen telopeptide regions while maintaining the integrity of the collagen triple-helical structure. Ultrasonic treatment ($p \leq 200$ W, $t \leq 15$ min) dispersed the collagen aggregates more evenly, and accelerated collagen self-assembly rate with a decreased but more homogeneous fibril diameter (82.78 ± 16.47–115.52 ± 19.51 nm) and D-periodicity lengths (62.1 ± 2.9–66.5 ± 1.8 nm) than that of the untreated collagen (119.15 ± 27.89 nm; 66.5 ± 1.8 nm). Meanwhile, ultrasonic treatment ($p \leq 200$ W, $t \leq 15$ min) decreased the viscoelasticity index and gel strength, enhancing thermal stability and promoting specific surface area and porosity of collagen fibril gels than that of the untreated collagen fibril gel. These results testified that collagen self-assembly behavior and collagen fibril gel properties can be regulated by ultrasonic treatment through multi-hierarchical structural alteration. This study provided a new approach for controlling in vitro collagen fibrillogenesis process so as to manufacture novel desirable collagen-based biomaterials with propitious performances for further valorization.

Keywords: ultrasonic treatment; collagen fibril; self-assembly behavior; collagen fibril gel; gel property

Citation: Liu, H.; Zhang, H.; Wang, K.; Qi, L.; Guo, Y.; Zhang, C.; Xu, Y. Impact of Ultrasonication on the Self-Assembly Behavior and Gel Properties of Bovine Bone Collagen I. *Molecules* **2023**, *28*, 3096. https://doi.org/10.3390/molecules28073096

Academic Editors: Shudong He, Wei Xu and Muhammad H. Aludatt

Received: 30 December 2022
Revised: 17 March 2023
Accepted: 27 March 2023
Published: 30 March 2023

1. Introduction

Collagen is the predominant structural protein of extracellular matrix (ECM), constituting approximately 1/3 of the total body proteins [1]. Type I collagen (CI) is the most abundant amongst the identified 29 collagen types (type I–XXIX), and also possess superior biodegradability and biocompatibility, which has now been widely utilized in food, cosmetics, biomedical and pharmaceutical industries [2,3]. Whether in vivo or in vitro under physiological conditions (specific pH, temperature, ionic strength, concentration, etc.), CI monomers can spontaneously aggregate side-by-side with each other in a quarter-staggered manner to self-assemble into microfibrils, fibrils, further into fibril bundles and, eventually, 3D collagen fibril gels [4]. Collagen fibrils and collagen fibril gels organize differently under different conditions through hydrogen bonding and hydrophobic–electrostatic interactions, and maintain optimized physicochemical properties and biological activity in vitro that are similar to those of native tissues [5]. As the most vital strength and viscoelasticity-related protein, the self-assembly behavior and fibril gel properties of CI make it responsible for the structural stability, mechanical properties, and application potentiality of collagen-based products in tissue engineering and traditional meat processing [6–8]. Therefore, different collagen self-assembly behavior and gel properties determine the different applied potency of collagen.

Many studies have found that collagen self-assembly behavior and fibril gel properties are strongly affected by relevant assembly conditions, such as collagen type/concentration, temperature, pH, ionic strength/species, surfactants, fibrillation inhibitors, external force field, etc. [4]. As an efficient and economic non-thermal assistant modification approach, low frequencies ultrasound (20–100 kHz, 1–1000 W/cm^2) has progressively been employed by food sectors as an alternative to conventional processes [9]. During ultrasonic treatment, electrical energy is converted into mechanical vibration energy, producing an acoustic cavitation effect and creating local high temperature and pressure on materials in a liquid system [10]. The acoustic cavitation generates sufficiently strong energy (10–100 kJ/mol) and physical shear forces that can break covalent and non-covalent bonds (e.g., hydrogen bonds and hydrophobic interactions) and ameliorating the physicochemical properties of proteins [11]. Li et al. (2018) reported that the D-spacing length and roughness of collagen fibrils from Qinchuan beef cattle tendons observed a 1.02% increment induced by ultrasonic treatment for 20 min (20, 28 and 40 kHz) using atomic force microscopy [12]. Yu et al. (2020) found that the ultrasonic cavitation and mechanical effect could accelerate the self-assembly process, reduce the fibrils size and alter digestion characteristics of pepsin-soluble collagen extracted from chicken leg skin [13]. Wan et al. (2021) studied the ultrastructure and mechanical properties of ultrasonic treatment collagen fibrils by atomic force microscopy [6].

In our previous study, ultrasonic treatment was applied to the chicken breast cartilage, and the result evinced that ultrasonication is an effective way to increase the yield of type II collagen and improve its physicochemical characteristics [8,9]. However, the potential effect of ultrasonication on the structural integrity, self-assembly behavior of the extracted collagen and the properties of the resulting collagen fibril gels has not yet been well-described, especially for bone collagen I. We have thus particularly recovered and characterized the attributes of bovine bone collagen I to valorize the low-value bovine bone byproducts as a high-quality alternative collagen source [14]. Bearing these in mind, this study aimed to further investigate the effect of ultrasonic treatment (power and time) on the self-assembly behavior and fibril gel properties of the extracted bovine bone collagen I. We expect that valuable data would be furnished for controlling the collagen fibril-formation process in vitro, and would lay a foundation for the fabrication of novel desirable bone collagen-based biomaterials and ECM analogues by ultrasonication.

2. Results and Discussion

2.1. Collagen Self-Assembly Behavior

2.1.1. Turbidity Assay

Collagen molecules possess all the information needed for fibrillogenesis so that it could self-assemble to be high-ordered fibrils when exposed to physiological conditions [2]. The collagen turbidimetric curve (Figure 1A) generally comprised three distinct stages to be a sigmoidal profile: lag phase (nucleation stage), growth phase (assembly stage) and plateau phase (equilibrium stage), indicating that the kinetics of collagen fibrillation followed a nucleation-dependent polymerization mechanism [15]. Nucleation occurred by the aggregation of collagen molecules to form dimers during the lag phase without detectable turbidimetric change, and the further aggregation of collagen molecules resulted in the formation of collagen fibrils with a steep sigmoidal increase in turbidity values in the growth phase [16].

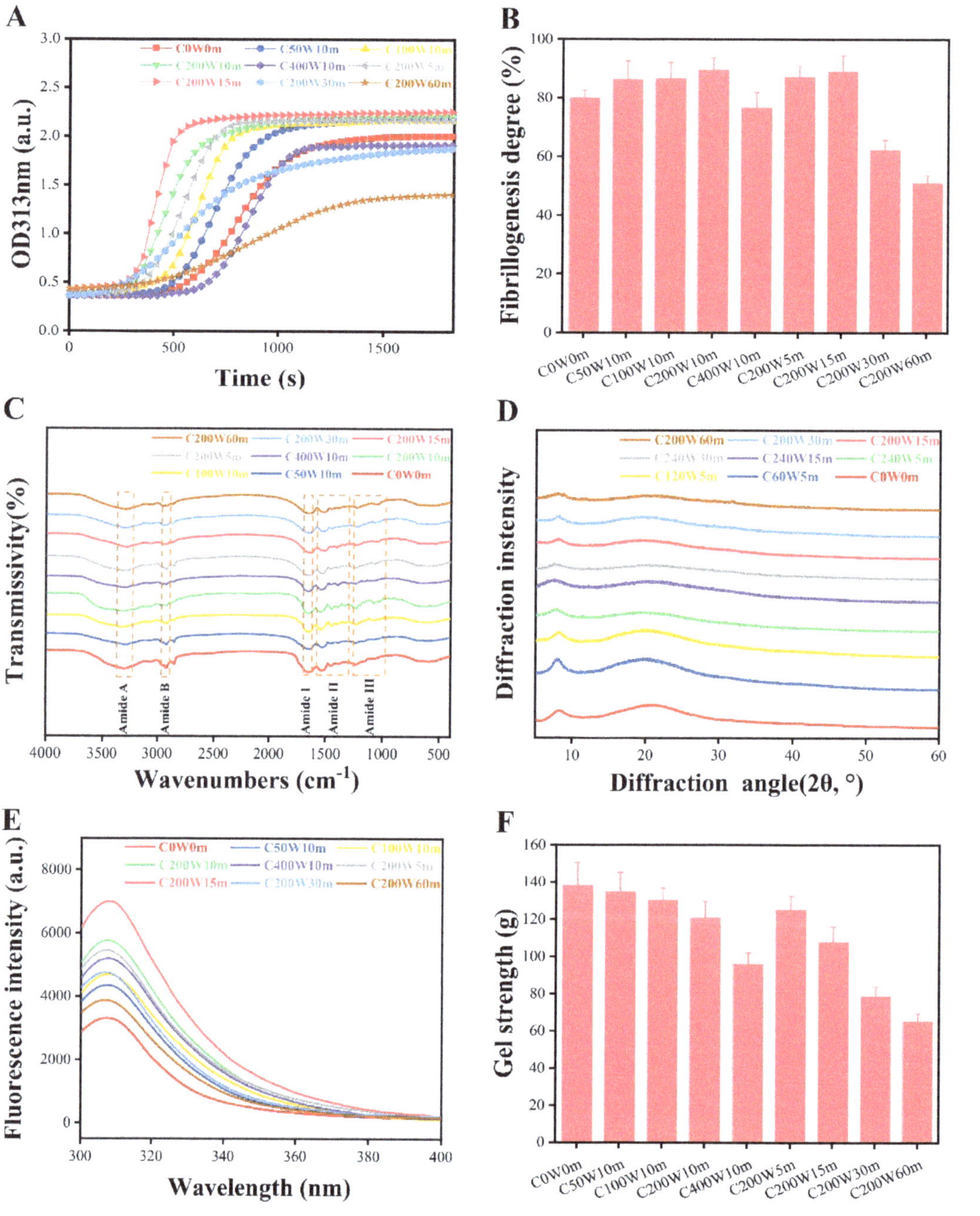

Figure 1. Turbidity assay (**A**), fibrillogenesis degree (**B**), fourier transform infrared spectroscopy (**C**), X-ray diffraction spectra (**D**) and fluorescence emission spectra (**E**) of collagen; gel strength of collagen fibril gels (**F**).

Under different ultrasonic treatment power (10 min), the turbidimetric curve of C0W0m exhibited a prolonged lag phase 0 s~590 s (590 s) and a growth phase of 590 s~1200 s (610 s). The turbidimetric curves of C50W10m (0 s~490 s, 490 s; 490 s~1015 s, 525 s), C100W10m (0 s~455 s, 455 s; 455 s~840 s, 385 s) and C200W10m (0 s~315 s, 315 s; 315 s~735 s, 420 s) displayed a shorter nucleation and assembly stage, but not for C400W10m (0 s~665 s, 665 s; 665 s~1085 s, 420 s) when compared to that of C0W0m. Results of the turbidity assay indicated that 50–200 W of ultrasonic treatment gradually increased the collagen fibrillogenesis rate, but 400 W decreased it. The fibrillogenesis degree (Figure 1B) also increased within 0–200 W (80.03–89.53%, 10 min) but decreased significantly at C400W10m (76.68%). These results showed that moderate ultrasonic treatment power accelerated collagen fibrillogenesis process and increased fibrillogenesis degree, with the optimal effect at 200 W. Under different ultrasonic treatment time (0–60 min, 200 W), turbidimetric curves of C200W5m (0 s~385 s, 385 s; 385 s~770 s, 385 s) and C200W15m (0 s~280 s, 280 s; 280 s~595 s, 315 s) showed a gradually increased lag phase and growth phase. However, the turbidimetric curves of C200W30m and C200W60m emerged with an unobvious sigmoidal profile and prolonged lag and growth phase, as well as a decreased equilibrium turbidity than that of C0W0m. In the plateau stage, collagen fibrils further assembled into supramolecular organization accompanied with an equilibrium of turbidity value [17].

These results suggested that moderate ultrasonic treatment ($p \leq 200$ W, t ≤ 15 min) collagen dispersed the collagen monomer homogeneously and increased the probability of collision and assembly, which could accelerate the nucleation stage so as to increase the fibrillogenesis rate [16]. However, high power and longtime ultrasonic treatment ($p > 200$ W, t > 15 min) weakened the collagen self-assembly ability, possibly because a negative impact was exerted on the structural integrity of collagen induced by cavitation effect [18].

2.1.2. Fibrillogenesis Degree

Fibrillogenesis is an entropy-driven process through hydrophobic and electrostatic interactions between the non-polar regions of adjacent molecules and hydrogen bonding between polar residues [19], as well as the specificity of molecular recognition [20]. These interactive forces minimize the surface area/volume ratio to be a circular cross-section by the loss of solvent molecules from collagen surface [21]. Collagen fibrillogenesis degree (Figure 1b) increased gradually within 0–200 W (80.03–89.53%, 10 min) and 5–15 min (87.26–89.03%, 200 W) but decreased significantly at C400W10m (76.68%), C200W30m (62.26%) and C200W60m (51.18%). These results suggested that moderate ultrasonic treatment ($p \leq 200$ W, t ≤ 15 min) was conducive to the fibrillogenesis process, but high power and longtime ultrasonic treatment ($p > 200$ W, t > 15 min) impaired it. Ultrasonication had possibly affected the molecular stretching and self-assembly process of collagen, thereby affecting the collagen fibrillogenesis degree [18,22]. The three-stage collagen fibrillogenesis process are generally accepted, yet the self-assembly mechanism is still lesser known [21].

2.2. Microstructure of Ultrasonic Treatment Collagen

2.2.1. SDS-PAGE

SDS-PAGE electrophoretic pattern (Figure 2) showed that all the collagen comprised of α1-, α2-, β- (dimers) and γ-chains (trimers), and the band intensity of α1-chains were approximately two-fold that of α2-chains (Table S2). These results testified that ultrasonic treatment did not destroy the covalent bond of collagen, and the structural integrity of its subunits (α1 and α2 chains) were maintained. The existence of high molecular weight bands (β- and γ-components) suggested that there existed various intra- and/or inter-molecular cross-linkages, which was conducive to maintain the thermal stability of collagen [11]. However, the diminished band intensity of β- and γ-chains was observed in C400W10m, C200W30m and C200W60m instead of C0W0m (Figure 2). These results suggested that high power and longtime ultrasonic treatment ($p > 200$ W, t > 15 min) had possibly induced an adverse impact on the structural and/or thermal stability of collagen.

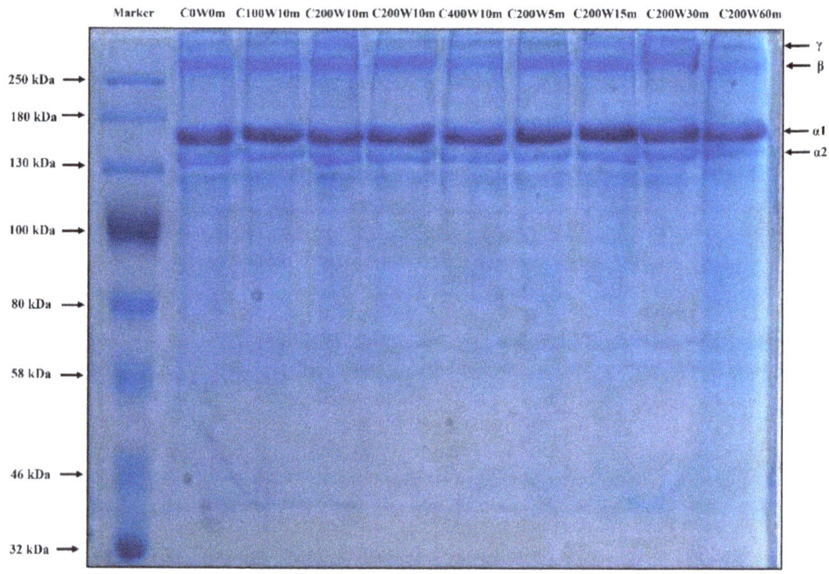

Figure 2. SDS-PAGE electrophoretic pattern of collagen.

2.2.2. Fourier Transform Infrared Spectroscopy (FTIR)

The FTIR spectra (Figure 1C) exhibited the collagen distinctive transmissivity peaks of amide A (3440–3400 cm^{-1}), B (2980–2850 cm^{-1}), I (1700–1600 cm^{-1}), II (1600–1550 cm^{-1}) and III (1360–1200 cm^{-1}) bands (Table S3), suggesting that all the collagen retained the triple-helical structure. The amide A and B bands are associated with free N-H stretching and asymmetric stretch vibrations of C-H and -NH_3^+, respectively [14]. The amide I band is a sensitive marker of secondary structure related to the stretching vibration of C=O groups coupled with N–H groups, but will shift to lower wavenumbers for the increased hydrogen bonds generated by the C=O groups with adjacent chains by stretching vibrations [8]. The amide II band is aroused by the N-H bending and CN stretching vibrations, while the amide III band is a weak absorption related to the N-H bending, C-N stretching, N-H bending vibrations and the CH_2 group wagging vibration in the glycine backbone proline side chains [23,24].

It was found that all the amide A band shifted to approximately 3300 cm^{-1}, indicating that the N-H groups were involved in hydrogen bonding with functional groups [25]. With the increase in ultrasonic treatment power and time, the amide B bands transfer from 2294 cm^{-1} to 2932 cm^{-1} due to the higher exposure of N-terminal free NH_3^+ groups of lysine residues induced by ultrasonic treatment [8,24]. The amide I band wavenumbers of collagen generally increased with the increased ultrasonic treatment power (1661–1651 cm^{-1}) and treatment time (1651–1634 cm^{-1}) than that of C0W0m (1653 cm^{-1}), proving that ultrasonic treatment decreased the molecular order degree and the number of hydrogen bonds of collagen. The amide II and III band wavenumbers of the ultrasonic treatment collagen (1520–1553 cm^{-1} and 1230–1240 cm^{-1}, respectively) were found higher than control (1518 cm^{-1} and 1230 cm^{-1}, respectively). These results suggested that non-ultrasonic treatment collagen had higher molecular order and more hydrogen bonds between adjacent α-chains than that of ultrasonic treatment collagen.

Curve-fitting analysis of amide I band of collagen (Table S4) also revealed that C0W0m had a higher content of α-helices + β-sheets (93.27%) than that of ultrasonic treatment collagen (56.96–86.85%), manifesting that ultrasonic treatment may exert adverse effects on the structural integrity of collagen. Furthermore, the peak intensity ratio ($A_{III}/A_{1\,450}$) represents a well-maintained and intact triple-helical structure if it is 1.05–1.14 [14]. $A_{III}/A_{1\,450}$ of all the ultrasonic/non-ultrasonic treatment collagen was 1.0487–1.1299 (Table S3), further proving that all the collagen maintained intact triple-helical structure. However, high power and longtime ultrasonic treatment ($p > 200$ W, $t > 15$ min) may weaken the structural integrity of collagen and induced multi-hierarchical structural differences, such as the number of hydrogen bonds, degree of molecular order, etc. These multi-hierarchical structural differences induced by ultrasonic treatment might greatly influence the physicochemical properties or biological activities of collagen.

2.2.3. X-ray Diffraction Spectra (XRD)

Two diffraction peaks at diffraction angles (2θ) aroused from the triple-helical structure and diffuse scatter caused by many structural layers of collagen, respectively, were observed in XRD spectra (Figure 1D). The result of the XRD spectra suggested that all the collagen retained their native triple-helical conformations and crystallinity [26]. The d values were calculated by the Bragg equation and represented the distance between crystal planes (Table S5). The d values of the first sharp peak (d1) and the second broad peak (d2) of collagen reflected the distance between the molecular chains and the distance between the skeletons of collagen, respectively [27]. The d1 values (related to the diameter of the triple-helix structure) of ultrasonic treatment collagen (1.107–1.174 nm) increased with the increased ultrasonic treatment power (50–400 W) and time (5–60 min) than that of C0W0m (1.088 nm). While the d2 values (related to the distance between amino acid residues along the helix of collagen) of ultrasonic treatment collagen (0.426–0.452 nm) increased within 50–200W (10 min) and 5–30 min (200W), they decreased at 400 W (10 min) and 60 min (200W) compared to that of C0W0m (0.425 nm). Consequently, the results of XRD showed that ultrasonic treatment did not damage the structural integrity but expanded the mainly intermolecular and then intramolecular distances of collagen molecules. The greater distance between the molecular chains of ultrasonic treatment collagen, such as C400W10m, C200W30m, and C200W60m, may make it more suitable as a drug delivery carrier than C0W0m [18].

2.2.4. Fluorescence Emission Spectra

The maximum fluorescence intensity of collagen fibrils (Figure 1E) increased with the increase of ultrasonic treatment power from 0 to 200 W, but decreased at 400 W (10 min). The results suggested that that the cavitation effect of ultrasonication induced the extension of the collagen molecules and fibrils, thereby exposing the inter- and intra-molecular tyrosine residues and increasing the fluorescence intensity. The maximum fluorescence intensity also increased with the extension of ultrasonic treatment time (5, 15 min) but decreased at 30 and 60 min (200 W). Tyrosine (intrinsic fluorophore) was present only at the N- and C-terminal telopeptide regions of collagen molecules and was responsible for initiating collagen fibrillation [28]. Therefore, it is speculated that the moderate ultrasonic cavitation effect reinforced the molecular stretching degree mainly through the exposure of tyrosine residues in the telopeptide regions of collagen molecules, thereby increasing the fluorescence intensity. However, high power and longtime ultrasonic treatment ($p > 200$ W, $t > 15$ min) may induce a smaller number of collagen self-assembled aggregates with a greater aggregation degree, thus reducing the fluorescence intensity.

2.2.5. Collagen Fibril Morphology (SEM/TEM)

SEM/TEM was adopted to observe the morphology of collagen fibrils. As was shown in Figure 3A1–I1, collagen fibrils were entangled into a dense and delicate fibrillar network with porous structure observed by SEM, possibly due to dehydration by freeze drying [18]. The average fibril diameters (Figure 3A2–I2) of C0W0m (119.15 ± 27.89 nm), C50W10m (115.52 ± 19.51 nm), C100W10m (113.63 ± 20.53 nm), C200W10m (105.35 ± 21.82 nm) and C400W10m (93.86 ± 22.05 nm) decreased gradually, and the average fibrils diameters of C200W5m (110.56±17.58 nm), C200W15m (96.33 ± 23.25 nm), C200W30m (88.38 ± 22.03 nm) and C200W60m (82.78 ± 16.47 nm) also exhibited a decreased trend. As for the characteristic collagen fibrils D-periodicity structure (Figure 4A–I), the average D-periodicity lengths of C0W0m (66.5 ± 1.8 nm), C50W10m (65.8 ± 2.2 nm), C100W10m (64.9 ± 2.6 nm), C200W10m (63.5 ± 1.6 nm) and C400W10m (62.1 ± 2.9 nm) decreased with the increase of ultrasonic treatment power (0-400 W), and the average D-periodicity lengths of C200W5m (64.1 ± 2.0 nm), C200W15m (62.8 ± 2.3 nm), C200W30m (61.7 ± 3.2 nm) and C200W60m (62.1 ± 3.6 nm) also exhibited a decreased trend.

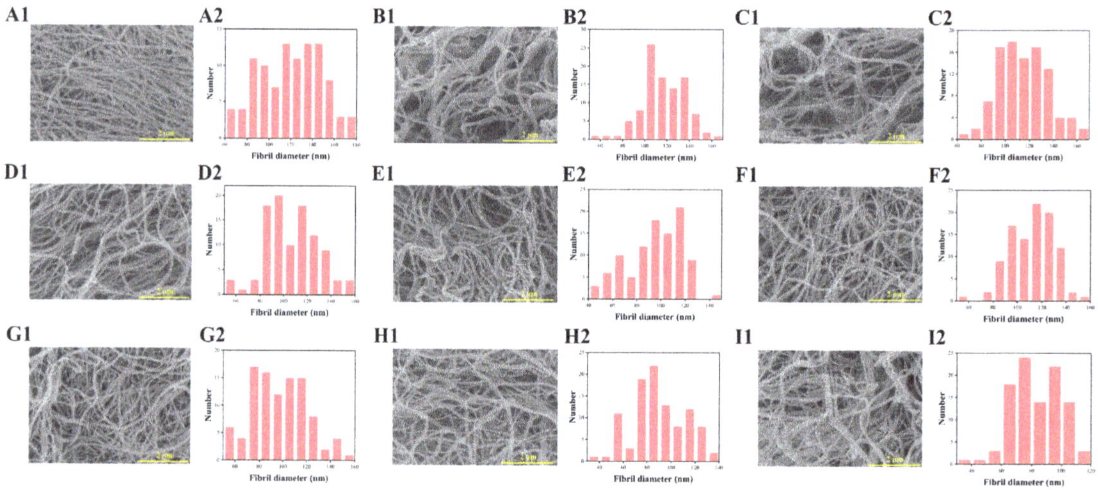

Figure 3. SEM images of collagen fibrils of C0W0m (**A1**), C50W10m (**B1**), C100W10m (**C1**), C200W10m (**D1**), C400W10m (**E1**), C200W5m (**F1**), C200W15m (**G1**), C200W30m (**H1**), C200W60m (**I1**). The fibril-diameter distribution of C0W0m (**A2**), C50W10m (**B2**), C100W10m (**C2**), C200W10m (**D2**), C400W10m (**E2**), C200W5m (**F2**), C200W15m (**G2**), C200W30m (**H2**) and C200W60m (**I2**).

Results of SEM/TEM demonstrated that ultrasonic treatment did not interrupt collagen fibrillogenesis and formation of D-periodicity, but diminished the fibril diameter and D-periodicity length of ultrasonic treatment collagen when compared with C0W0m. Furthermore, the fibril-diameter distribution of ultrasonic treatment collagen was more homogeneous. Previous studies had concluded that ultrasonic treatment mainly accelerate the collagen self-assembly process at the nucleation stage through increasing collagen molecular interaction [7]. Meanwhile, ultrasonic treatment dispersed trivalent phosphate ions in solution evenly through the cavitation effect to widely bind on the high excess positively charged regions of collagen molecules to accelerate the collagen self-assembly process [7]. However, collagen fibrillogenesis became a spontaneous and orderly molecular arrangement process after the formation of collagen fibril nucleus, and thereby ultrasonic treatment has no significant effect on the kinetics of self-assembly and the equilibrium stage.

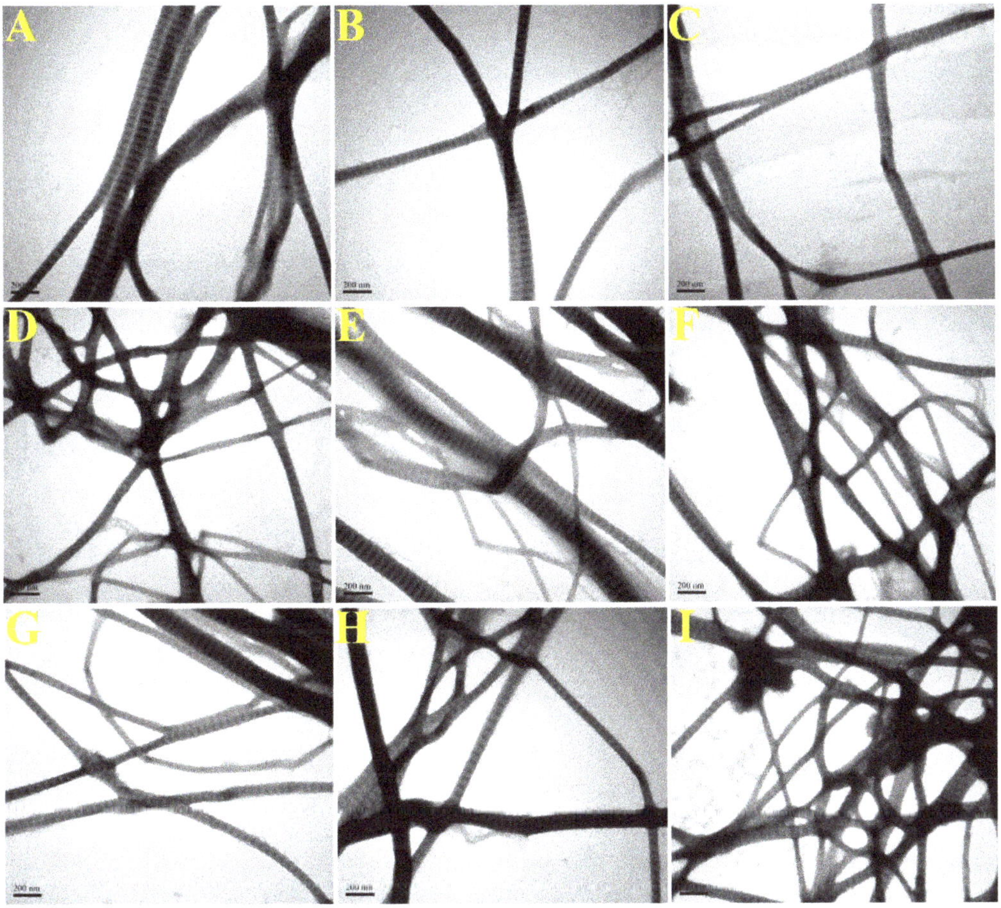

Figure 4. TEM images of collagen fibrils of C0W0m (**A**), C50W10m (**B**), C100W10m (**C**), C200W10m (**D**), C400W10m (**E**), C200W5m (**F**), C200W15m (**G**), C200W30m (**H**) and C200W60m (**I**).

The fibril diameters and D-periodicity width of ultrasonic treatment collagen exhibited lower heterogeneity than that of control (C0W0m). In this regard, the collagen monomer molecules accumulated randomly into microfibrils at the nucleation stage, and then the microfibrils further assembled into fibrils, resulting in the heterogeneous distribution of microfibril and fibril diameters [13]. Ultrasonic treatment may disperse the collagen monomers more evenly through cavitation effect, increasing the number of collagen microfibrils formed in the nucleation stage. Therefore, the increased number of collagen microfibrils will decrease the average fibril diameter and fibril diameter heterogeneity when the number of total collagen molecules is constant. Mechanisms of D-periodicity formation are still not well-understood, but ultrasonic treatment may affect the length of gap areas and overlapping areas through variable molecular arrangement, resulting in the smaller D-periodicities according to the 1/4 staggered model of Schmitt hypothesis [19,29]. The collagen fibrils with smaller diameter interact more affine with ECM components, facilitating elasticity and decreased plastic deformation [2]. The uniform morphology of collagen fibrils (diameter size distributions, network structure, etc.) was a significant favorable property for biomedical and pharmaceutical applications [30].

2.3. *Thermal Stability of Collagen Fibril Gels*

2.3.1. Differential Scanning Calorimetry (DSC)

The DSC curves of all collagen fibril gels (Figure 5A) showed a typical thermal denaturation peak induced by heating. DSC spectra showed that the melting temperature (Tm) of collagen fibril gels increased from 50.5 °C (C0W0m) to 52.1 °C (C200W10m) with the ultrasonic treatment power increased from 0 W to 200 W but decreased to 47.8 °C at 400 W (C400W10m); Tm decreased from 53.5 °C (C200W5m) to 46.3 °C (C200W60m) with the ultrasonic treatment time increased from 5 min to 60 min. The enthalpy (ΔH) of collagen fibril gels decreased from 3.19 J/g (C0W0m) to 1.39 J/g (C400W10m) with the ultrasonic treatment power increased from 0 W to 400 W (10 min). ΔH values also showed a descending trend from 2.78 J/g (C200W15m) to 1.86 J/g (C200W60m) with the ultrasonic treatment time increased from 5 min to 60 min (200 W). Tm values depended on the nucleation stage of collagen fibrillogenesis process, and moderate ultrasonic treatment accelerated the nucleation stage to increase the fibrillogenesis degree and thereby strengthen the thermal stability [4]. The amount of intermolecular and intramolecular hydrogen bonds of collagen determined ΔH values. Ultrasonic treatment decreased the number of formed hydrogen bonds and thereby induced a lower ΔH values [8]. These results suggested that ultrasonic treatment will enhance the thermal stability and thus greatly expand the applied potency of collagen fibrils and collagen fibril gels, which may be related to the change in collagen conformation induced by ultrasonic treatment.

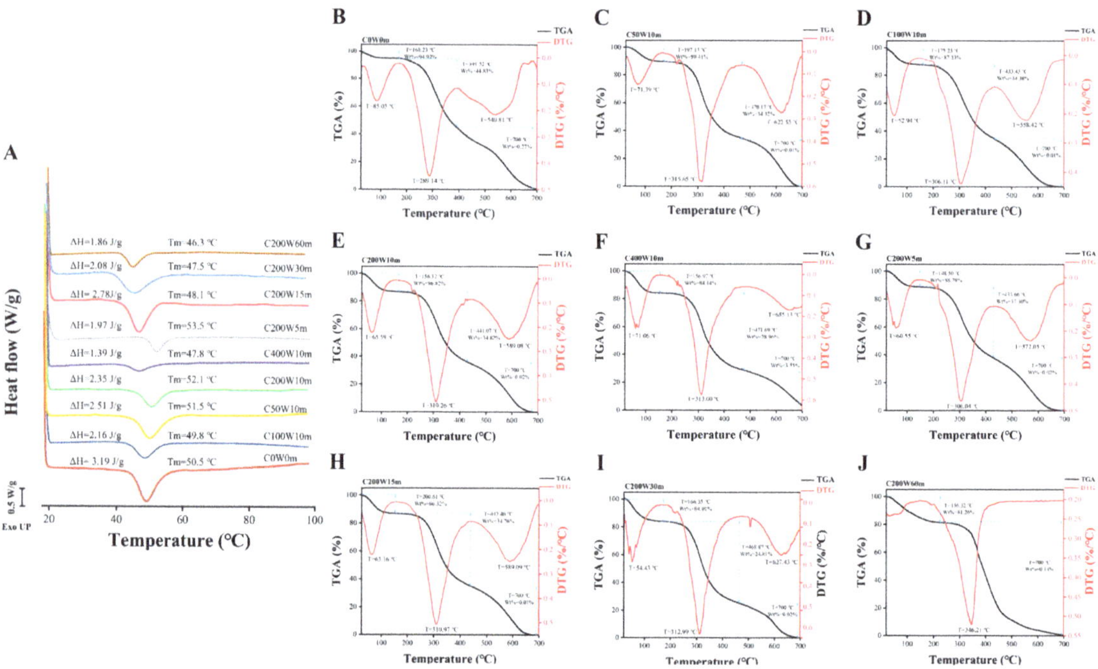

Figure 5. Differential scanning calorimetry curves of collagen fibril gels at 20–100 °C (**A**); thermogravimetric curves (TGA) and its first derivatives (DTG) of C0W0m (**B**), C50W10m (**C**), C100W10m (**D**), C200W10m (**E**), C400W10m (**F**), C200W5m (**G**), C200W15m (**H**), C200W30m (**I**) and C200W60m (**J**).

2.3.2. Thermo-Gravimetric Analysis (TGA)

The further thermal destruction of collagen fibril gels in a large temperature range was characterized by thermogravimetric curves (TGA) and its first derivatives (DTG) (Figure 5B–J) as well as weight loss per 100 °C (Table S6). The three peaks observed in all DTG plots revealed a three-stage thermal destruction of collagen fibril gels (Figure 5B–J). An initial thermal transitional change was observed at 52.94–200.61 °C, which was attributed to the removal of free water [24]. A maximum percentage transitional weight loss of 5.08% at 168.23 °C (C0W0m) to 15.86% at 156.97 °C (C400W10m) and 11.22% at 148.50 °C (C200W5m) to 18.74% at 156.32 °C (C200W60m) demonstrated that ultrasonic treatment accelerated the loss of free water. A greater water loss of ultrasonic treatment collagen suggested that they were more permeable and thereby suitable for using as carrier or catalyst [24]. In the second stage, collagen binding water, and then small molecular thermal degradation products, such as peptides and amino acids, were released [31]. C0W0m (55.17% at 391.52 °C) exhibited a minimum percentage of transitional weight loss and transitional temperature than that of ultrasonic treatment collagen (62.70–75.19% and 433.43–471.69 °C), which may be due to the fewer hydrogen bonds in the collagen induced by ultrasonic treatment. In the third stage, high temperature further decomposed peptides and amino acids into CO_2, CO and NO via deamination and dehydration [31], and all the collagen fibril gels completely decomposed. These results suggested that moderate ultrasonic treatment enhanced the impermeability of collagen fibril gels, but high power and/or longtime ultrasonic treatment ($p > 200$ W, $t > 15$ min) may weaken the thermal stability of collagen fibril gels in a large temperature range.

2.4. *Viscoelasticity of Collagen Fibril Gels (Dynamic Frequency Sweep Test)*

The viscoelastic properties of collagen fibril gels were evaluated by storage modulus (G', describes the elasticity behavior) and the loss modulus (G'', describes the dissipated energy as a characteristic viscosity) obtained from dynamic frequency sweep tests [32,33]. G' represents the structure stiffness to deform with the impact of external force; it depends on the number and strength of the secondary bonds, while G'' modulus reflects the energy loss induced by collagen intra- or inter-molecular stretching when the external force changes [9]. As was shown in Figure 6A,B, the G' and G'' moduli of all gels increased when the shear frequency increased from 0.01 to 10 Hz, while the G' modulus was much larger than the corresponding G'', which was the typical rheological attribute of the gel structure. The G' and G'' moduli of ultrasonic treatment collagen generally decreased with increased ultrasonic treatment power and time than that of C0W0m, indicating that ultrasonic treatment could have impaired the viscoelasticity of collagen fibril gels during self-assembly [34,35].

Ultrasonic treatment may alter the entanglement networks between collagen macromolecules as well as configurational rearrangements by short-range relaxation times (between entanglements) and long-range relaxation times (beyond entanglements), thereby endowing the reduced dynamic viscoelastic behavior of collagen fibril gels [35]. Tan δ (G''/G', defined as tangent loss angle) represented the mechanical loss of polymer in oscillatory motion, reflecting the network structure of gel. The smaller the tan δ value, especially when $\tan\delta < 0.3$, the better the gel network structure formed by the protein [36,37]. It was found that all the ultrasonic treatment collagen showed a smaller tan δ value ($\tan\delta < 0.3$, Figure 6C) than that of C0W0m, except for C200W60m ($\tan\delta > 0.3$), suggesting that ultrasonic treatment was conducive to the formation of collagen fibril gel network.

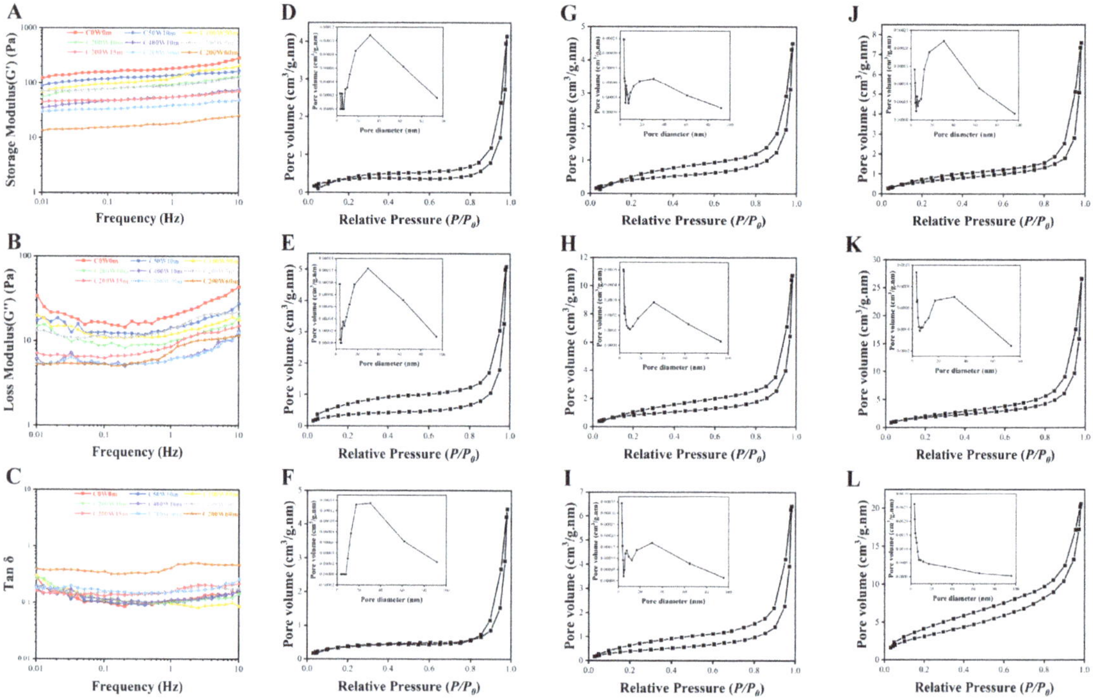

Figure 6. Storage modulus (G′), (**A**), loss modulus (G′′), (**B**) and loss tangent (tan δ), (**C**) of collagen fibril gels; N_2 adsorption isotherms and BJH pore size distribution curves of C0W0m (**D**), C50W10m (**E**), C100W10m (**F**), C200W10m (**G**), C400W10m (**H**), C200W5m (**I**), C200W15m (**J**), C200W30m (**K**) and C200W60m (**L**).

2.5. Gel Strength

Gel strength represents the transition capacity of colloidal sol into polymer gel. As was shown in Figure 1F, a gradual descending trend of collagen fibril gel strength along with increases in ultrasonic treatment power and time were observed. Ultrasonic treatment during the nucleation stage of collagen fibrillogenesis endowed collagen molecules a higher stretching degree, and the increased hydrophobicity weakened the collagen–water interactions. Therefore, ultrasonic treatment created more tenuous and looser collagen fibril gels compared with C0W0m, and then the collagen fibrillar gel strength was diminished to be "softer" gels. Collagen gels are widely used as biomedical materials in the fields of scaffolds, delivery matrices, catalysts, and injectable carriers. Wherein the biomedical scaffolds require sufficient strength and elasticity, the soft and low-viscosity are more suitable for delivery matrices [7]. In this study, ultrasonic treatment affected the nucleation stage of the collagen fibrillogenesis process, and then reduced the collagen fibril gel strength, which greatly expanded the applied potency of collagen fibril gel.

2.6. Nitrogen Adsorption of Collagen Fibril Gels

The nitrogen adsorption isotherm and pore size distribution of collagen fibril gels were shown in Figure 6D–L, and the hysteresis loops suggested they were all type IV isotherms [18]. The specific surface area of collagen fibril gels calculated by the BET method increased from 1.380 m^2/g to 3.179 m^2/g when ultrasonic treatment power was 0~400 W (10 min), and increased as well from 1.529 to 12.013 m^2/g when ultrasonic treatment time was 5~60 min (200 W). BJH analysis (Table S7) showed that the total pore volumes of collagen fibril gels generally showed an increased trend from 0.0064 cm^3/g to 0.0166 cm^3/g, and their average pore diameter decreased from 22.466 nm to 16.916 nm when the ultrasonic treatment power was 0~400 W (10 min). The total pore volumes and average pore diameter of collagen fibril gels increased from 0.0099 cm^3/g to 0.0318 cm^3/g and decreased from 25.917 nm to 10.581 nm, respectively, when the ultrasonic treatment time was 5~60 min (200 W). BJH analysis showed that high power and longtime ultrasonic treatment created mesoporous collagen fibril gels with larger specific surface area and pore volume but more homogeneous pore diameter than natural fibrils (specific surface area 1~5 m^2/g). Consequently, these results indicated that ultrasonic treatment may be a potential method to manufacture homogeneous porous collagen scaffolds and biomaterials as catalyst and carrier.

3. Materials and Methods

3.1. Raw Materials and Chemical Reagents

Freeze-dried bovine bone collagen was prepared by our previous study, which was characterized as type I collagen with the molecular form of [α1(I)]2α2(I) [14]. All other reagents and chemicals used in this study were of analytical grade and obtained from Sinopharm Chemical Reagent Co., Ltd. (Shanghai, China).

3.2. Ultrasonic Treatment on Collagen

Bovine bone collagen I was dialyzed in a phosphate buffer (200 mM, pH 7.4) at 4 °C for 48 h after being stirred to dissolve in 10 mM HCl solution (pH 2.0) at a concentration of 5 mg/mL. Subsequently, 200 mL of collagen solutions was incubated with different power and duration of ultrasonication in a glass beaker using an ultrasonic processor (VCX 750, Sonics & Materials Inc., Newtown, CT, USA) equipped with a high-grade titanium alloy probe. The output power was adjustable (0–100%) with a rated power of 750 W (20 kHz) and emitting surface of 13.0 mm diameter. The temperature of the glass was maintained at 30 °C by a temperature-controlled steel jacket through a circulating water bath (30 °C), and the ultrasonic processor worked in a pulse mode of 2 s acting and 3 s resting time to avoid overheating [9]. The probe was immersed in the solution to a depth of 3 cm from the bottom during the ultrasonic treatment process. The collagen solutions were first treated with different ultrasonic treatment power (0–400 W) for 10 min to investigate the effect of ultrasonic treatment power on collagen self-assembly behavior; then, the optimal ultrasonic treatment power (200 W, according to fibrillogenesis rate and fibrillogenesis degree) was selected to study the effect of ultrasonic treatment time (0–60 min) on the collagen self-assembly behavior. The ultrasonic treatment collagen was termed as C0W0m, C50W10m, C100W10m, C200W10m, C400W10m, C200W5m, C200W15m, C200W30m, and C200W60m, respectively (Table 1). After the ultrasonic treatment process, all the collagen solutions (200 mL) were further incubated at 30 °C to 60 min, and 100 mL of the collagen solutions were freeze-dried for further utilization. The actual output ultrasonic treatment power (W) was calculated using the following Equation (1), and the ultrasonic treatment intensity (W/cm^2, Table 1) equals to the output power divided by the emitting surface (1.327 cm^2) [38].

$$P = m \times C_p \times \partial T/\partial t \tag{1}$$

where P is the power output, m is the mass of the sonicated liquid (g), C_p is the specific calorific capacity of the fluid (J/g °C), and $\partial T/\partial t$ is the temperature change rate of the solvent over a 3 min period. In this experiment, the ultrasonication intensity was 0.62–38.47 W/cm^2 calculated by a calorimetric study of ultrasonic processor VCX 750 (Table 1).

Table 1. Calorimetric study on VCX 750 ultrasonic probe and ultrasonication conditions for collagen samples.

Samples	Power Setting (W)	Actual Power Output (W)	Ultrasonic Time at 30 °C (min)	Ultrasonication Intensity (W/cm^2) [a]	Incubation Time at 30 °C in the Absence of Ultrasonic (min)
C0W0m	0	0	0	0	60
C50W10m	50	0.82	10	0.62	50
C100W10m	100	1.56	10	1.15	50
C200W10m	200	8.32	10	6.27	50
C400W10m	400	51.05	10	38.47	50
C200W5m	200	8.32	5	6.27	55
C200W15m	200	8.32	15	6.27	45
C200W30m	200	8.32	30	6.27	30
C200W60m	200	8.32	60	6.27	0

[a]: Ultrasonic intensity is equal to the power output measured by calorimetry divided by the area of the emitting surface (1.327 cm^2).

3.3. *Turbidity Assay*

Fibrillogenesis process of bone collagen I was monitored by the turbidity assay according to Tian et al. (2021) [4], because the turbidity of collagen solutions changes during fibrillogenesis in simulated body fluid solution (SBF). The obtained ultrasonic treatment collagen solutions in Section 3.2 (4 mL) were mixed with 1 mL of 10× *g* concentrated SBF (Table S1) at 4 °C, and the solutions were adjusted to pH 7.4 with 2 M NaOH to prepare the collagen stock solutions. The collagen stock solutions were transferred into a quartz cuvette (1 cm) and then incubated in a spectrophotometer (PE Lambda 25, Perkin Elmer, Waltham, MA, USA) at 30 °C for 60 min. The absorbance of collagen mixtures at 313 nm was recorded every 35 s by a UV spectrophotometer (PE Lambda 25, Perkin Elmer, Waltham, MA, USA) during fibrillogenesis.

3.4. *Fibrillogenesis Degree*

Fibrillogenesis degree of collagen is defined as the percent of collagen molecules that reassemble into the fibrils [18]. The fibrillogenesis process was conducted as described in Section 3.3 at 30 °C for 12 h, then the collagen mixtures were centrifuged at 18,000× g (20 °C) for 20 min. The protein content in supernatant was measured by a Lowry protein quantification assay kit, and the collagen fibrillogenesis degree was calculated using Equation (2):

$$Y(\%) = \frac{C_0 - C}{C_0} \times 100\% \quad (2)$$

where Y is the collagen fibrillogenesis degree (%), C_0 (mg/mL) and C (mg/mL) is the protein concentration before and after collagen fibrillogenesis.

3.5. *Microstructure of Ultrasonic Treatment Collagen*

3.5.1. Sodium Dodecyl Sulphate Polyacrylamide Gel Electrophoresis (SDS-PAGE)

SDS-PAGE was performed according to the method of Thuy et al. (2014) [39] with slight modifications using BIO-RAD Mini-PROTEAN gels system (Bio-Rad Laboratories, Hercules, CA, USA). The lyophilized collagen was dissolved in 0.1 M acetic acid into a concentration of 5 mg/mL. Then, collagen solutions were mixed with sample buffer (0.5 M Tris–HCl, pH 6.8, containing 10% (*w*/*v*) SDS, 20% (*v*/*v*) glycerol and 0.02% bromophenol blue) at a ratio of 1:1 (*v*/*v*). Each sample (approximately 10 µg) was loaded onto the gels and the electrophoresis was conducted on 80 V for 4% stacking gel and 110 V for 7.5% resolving gel. Gels were stained with 0.1% (*v*/*v*) Coomassie brilliant blue R250 and then decolorized by the decolorizing solution (30% methanol + 10% acetic acid, *v*/*v*). Gels were preserved in 7% (*v*/*v*) acetic acid until they were photographed by the Alpha Ease FC gel imaging system (Alpha Innotech, San Leandro, CA, USA). Quantitative analysis of collagen band intensity was performed using ImageJ 1.8.0 software (National Institute of Mental Health, Bethesda, MD, USA).

3.5.2. Fourier Transform Infrared Spectroscopy (FT-IR)

FTIR spectra of lyophilized collagen were recorded in the range of 4000–400 cm^{-1} with 1 cm^{-1} resolution for a single scan using a Nicolet iS10 FT-IR spectrometer (Thermo Fisher Scientific Inc., Madison, WI, USA) according to the procedure depicted by Kang D et al. (2016) [40]. FT-IR spectra data were analyzed with OMNIC software (v8.20, Thermo Nicolet, Madison, WI, USA).

3.5.3. X-ray Diffraction (XRD)

XRD spectra of lyophilized collagen were obtained following the method described by Pezeshk S et al. (2022) [41] with slight modification using an XRD instrument (XRD-6000, Shimadzu, Kyoto, Japan) with copper Kα as a source of X-rays: scanning range 5–60° (2θ); scanning speed 2°/min. The minimum value of the repeated spacings (d values) was calculated by the Bragg equation, as in Equation (3):

$$d\left(\text{Å}\right) = \lambda/2\sin\theta \quad (3)$$

where λ is the wavelength of copper Kα X-ray (1.54 Å) and θ is the Bragg diffraction angle.

3.5.4. Fluorescence Emission Spectra

Endogenous fluorescence is produced by aromatic amino acids (tryptophan, tyrosine, phenylalanine) for their benzene ring or conjugated double bond structure [42]. The content of tryptophan and phenylalanine are low in collagen, and thus, tyrosine was selected as the endogenous fluorescence of collagen to analyze the effect of ultrasonic treatment on collagen conformation. According to the method described by Soumya N. et al. (2021) [15], ultrasonic treatment collagen solutions in Section 3.2 were diluted with 10 mM HCl to a concentration of 0.5 mg/mL. The fluorescence spectrum of collagen solutions was recorded using a fluorescence spectrophotometer (F-4600, Hitachi, Tokyo, Japan). The excitation wavelength was 280 nm, the emission wavelength 300–400 nm, the excitation and emission spectral slit 5 nm, while the scanning rate was 1200 nm/min.

3.6. Microscopic Structure of Collagen Fibrils (SEM/TEM)

The morphology of collagen fibrils was observed using scanning electron microscope (SEM; S4800, Hitachi, Tokyo, Japan) according to the procedures described by Ran et al. (2020) [16]. Collagen stock solutions (1 mg/mL) were prepared as described in Section 3.2, and 50 μL of collagen stock solutions were incubated overnight at 30 °C in clean slides. The treated samples were rinsed carefully with deionized water for several times and fixed with 2.5% (*v*/*v*) glutaraldehyde for 2 h, then dehydrated with 50% (*v*/*v*) ethanol and freeze-drying with a freeze dryer (SCIENTZ-10ND, Ningbo Xinzhi Biotechnology Co. ltd., Ningbo, China). The dried samples were observed by SEM at an accelerating voltage of 5 kV, and the diameters of 100 fibrils were analyzed with ImageJ software (v1.51, National Institutes of Health, Bethesda, MD, USA).

Observation of the collagen fibrils by transmission electron microscopy (TEM) were performed according to the method of Liu et al. (2014) [29] with slight modifications. Collagen stock solutions were prepared the same as Section 3.2. One drop of the solution was loaded on a copper grid with 200 mesh size and stained with 1% (*w*/*v*) phosphotungstic acid, and then then washed with deionized water and air dried. The collagen fibrils were observed using a JEM-2100 electron microscopy (JEOL Ltd., Tokyo, Japan) with an accelerating voltage of 200 kV.

3.7. Preparation of Collagen Fibril Gels

The ultrasonic treatment collagen solutions (40 mL) were mixed with 10 mL of 10× *g* concentrated SBF at 4 °C, and the solutions were adjusted to pH 7.4 with 2 M NaOH. The collagen mixture solutions were incubated for 12 h at 25 °C to prepare collagen fibril gels, which were then incubated for 12 h and kept at 4 °C for further analyses [4].

3.8. Thermal Stability of Collagen Fibril Gels

3.8.1. Differential Scanning Calorimetry (DSC)

The melting temperature (Tm) of collagen fibril gels were quantified using Q2000 Series DSC (TA Instruments, Inc., New Castle, DE, USA). Samples (approximately 5 mg) were accurately weighted and sealed into the aluminum pan, and an empty aluminum pan was used as the reference. The pan was equilibrated at a rate of 5 °C/min and heated from 10 to 100 °C [9].

3.8.2. Thermogravimetric Analysis (TGA)

The thermal properties of the collagen fibril gels were evaluated using a TG/DTA instrument (Pyris Diamond 6000 TG/DTA, PerkinElmer, USA) from 40–700 °C at 10 °C/min in a nitrogen atmosphere, and the reported data were averages of three scans [43].

3.9. Collagen Fibril Gel Strength

The strength of collagen fibril gels (height 30 mm × diameter 20 mm) was measured following the method of Jiang et al. (2016) [7] using a TA-XT2i texture analyzer (Stable Micro Systems, Surrey, UK). The strength of collagen fibril gels was determined using a cylindrical probe (P/0.5) at a constant velocity of 1.0 mm/s, with five samples for each determination.

3.10. Viscoelasticity Properties of Collagen Fibril Gels

The dynamic viscoelasticity of collagen fibril gels was measured following the procedure of Tian et al. (2021) [4] through dynamic frequency sweep tests from 0.01 to 10 Hz at 25 °C using a rheometer (MCR 302, Anton Paar, Austria). The rheometer was equipped with a temperature-controlled stainless-steel cone/plate geometry (0.5° cone angle, 60 mm cone diameter, 57 μm gap). Storage modulus (G'), loss modulus (G''), and loss tangent (tan δ) were recorded.

3.11. Nitrogen Adsorption (BET) of Collagen Fibril Gels

The specific surface areas, pore size and pore volume of collagen fibril gels were determined by a surface area analyzer (Autosorb-6B, Quantachrome Instruments, Boynton Beach, FL, USA) based on the Brunauer–Emmett–Teller (BET) method and the Barrett–Joyner–Halenda (BJH) method, following the procedure described by Liao et al. (2018) [18]. The pore structure information of collagen fibril gels was calculated from the desorption branch.

3.12. Statistical Analysis

All experiments were made in triplicate and results were presented as mean ± standard deviation (SD). Statistical analyses were analyzed using ANOVA's test ($p < 0.05$) by SPSS 26.0 (IBM SPSS Statistics, Ehningen, Germany) and all the figures were processed by Origin 9.0 software (OriginLab Co., Northampton, MA, USA).

4. Conclusions

Results of SDS-PAGE, FT-IR, XRD and fluorescence emission spectra showed that ultrasonic treatment did not damage the integrity of collagen triple-helical structure, but induced multi-hierarchical structural alteration and dispersed the collagen aggregates evenly. Proper ultrasonic treatment ($p \leq 200$ W, t ≤ 15 min) accelerated collagen self-assembly rates and gained a decreased but more homogeneous fibril diameter and D-periodicity lengths. Proper ultrasonic treatment ($p \leq 200$ W, t ≤ 15 min) also impaired the viscoelasticity properties and enhanced the thermal stability of collagen fibril gels, promoting specific surface area and porosity of collagen fibril gels while creating a softer gel. Consequently, this study may contribute a new avenue for controlling collagen fibrillogenesis in vitro and ameliorating collagen fibril gel properties to manufacture novel collagen biomaterials with desirable performances for further valorization. However, the optimum ultrasonic treatment power and duration should be carefully evaluated depending on the requirement.

Supplementary Materials: The following supporting information can be downloaded at: https://www.mdpi.com/article/10.3390/molecules28073096/s1, Table S1: Components of simulated body fluid solution (mM); Table S2: SDS-PAGE band intensities of collagen; Table S3: Fourier transform-infrared spectra peak wavenumber and assignment of collagen; Table S4: Curve-fitting analysis of amide I band of collagen; Table S5: D values calculated by the Bragg equation of the two x-ray diffraction peaks of collagen; Table S6: Thermal analysis of collagen fibril gels under N_2 air atmosphere; Table S7: N_2 adsorption analysis of collagen fibril gels.

Author Contributions: Conceptualization, H.L.; methodology, H.L.; software, H.L.; validation, K.W., L.Q. and Y.X.; formal analysis, Y.G.; investigation, K.W., L.Q. and Y.G.; resources, C.Z.; data curation, H.L.; writing—original draft preparation, H.L.; writing—review and editing, H.Z.; visualization, Y.G.; supervision, C.Z. and Y.X.; project administration, C.Z.; funding acquisition, Y.G. and C.Z. All authors have read and agreed to the published version of the manuscript.

Funding: This research was funded by Technical Breakthrough Projects of Inner Mongolia Science and Technology Program (No. 2022JBGS0007), National Agricultural Science and Technology Innovation Project (No. CAAS-ASTIP-2023).

Institutional Review Board Statement: Not applicable.

Informed Consent Statement: Not applicable.

Data Availability Statement: Not applicable.

Acknowledgments: We thank all the colleagues who contributed to this research.

Conflicts of Interest: The authors declare no conflict of interest.

References

1. Qin, L.; Bi, J.R.; Li, D.M.; Dong, M.; Zhao, Z.Y.; Dong, X.P.; Zhou, D.Y.; Zhu, B.W. Unfolding/refolding study on collagen from sea cucumber based on 2D fourier transform infrared spectroscopy. *Molecules* **2016**, *21*, 1546. [CrossRef] [PubMed]
2. Gunasekaran, D.; Thada, R.; Jeyakumar, G.F.S.; Manimegalai, N.P.; Shanmugam, G.; Sivagnanam, U.T. Physicochemical characterization and self-assembly of human amniotic membrane and umbilical cord collagen: A comparative study. *Int. J. Biol. Macromol.* **2020**, *165*, 2920–2933. [CrossRef] [PubMed]
3. Zou, Y.E.; Xu, P.; Li, P.; Cai, P.; Zhang, M.; Sun, Z.L.; Sun, C.; Xu, W.M.; Wang, D.Y. Effect of ultrasound pre-treatment on the characterization and properties of collagen extracted from soft-shelled turtle (Pelodiscus sinensis). *LWT Food Sci. Technol.* **2017**, *82*, 72–81. [CrossRef]
4. Tian, H.H.; Ren, Z.Y.; Shi, L.F.; Hao, G.X.; Chen, J.; Weng, W.Y. Self-assembly characterization of tilapia skin collagen in simulated body fluid with different salt concentrations. *Process Biochem.* **2021**, *108*, 153–160. [CrossRef]
5. Ju, H.Y.; Liu, X.Y.; Zhang, G.; Liu, D.Z.; Yang, Y.S. Comparison of the structural characteristics of native collagen fibrils derived from bovine tendons using two different methods: Modified acid-solubilized and pepsin-aided extraction. *Materials* **2020**, *13*, 358. [CrossRef]
6. Wan, Y.F.; Gao, Y.F.; Shao, J.H.; Tumarbekova, A.; Zhang, D.Q.; Zhu, J. Effects of ultrasound and thermal treatment on the ultrastructure of collagen fibers from bovine tendon using atomic force microscopy. *Food Chem.* **2021**, *347*, 128985. [CrossRef]

7. Jiang, Y.; Wang, H.B.; Deng, M.X.; Wang, Z.W.; Zhang, J.T.; Wang, H.Y.; Zhang, H.J. Effect of ultrasonication on the fibril-formation and gel properties of collagen from grass carp skin. *Mater. Sci. Eng. C* **2016**, *59*, 1038–1046. [CrossRef]
8. Akram, A.N.; Zhang, C.H. Extraction of collagen-II with pepsin and ultrasound treatment from chicken sternal cartilage; physicochemical and functional properties. *Ultrason Sonochem.* **2020**, *64*, 105053. [CrossRef] [PubMed]
9. Akram, A.N.; Zhang, C.H. Effect of ultrasonication on the yield, functional and physicochemical characteristics of collagen-II from chicken sternal cartilage. *Food Chem.* **2020**, *307*, 125544. [CrossRef]
10. Ye, Y.K.; Dai, S.Q.; Zhang, H.Y.; He, S.D.; Hu, W.W.; Cao, X.D.; Wei, Z.J. Ultrasound-Assisted Preparation of Maillard Reaction Products Derived from Hydrolyzed Soybean Meal with Meaty Flavor in an Oil-In-Water System. *Molecules* **2022**, *27*, 7236. [CrossRef]
11. Ata, O.; Kumcuoglu, S.; Tavman, S. Effects of sonication on the extraction of pepsin-soluble collagens from lamb feet and product characterization. *LWT Food Sci. Technol.* **2022**, *159*, 113253. [CrossRef]
12. Li, G.X.; Wan, Y.F.; Song, X.; Wang, Y.; Zan, L.S.; Zhu, J. Effects of various processing methods on the ultrastructure of tendon collagen fibrils from Qinchuan beef cattle observed with atomic force microscopy. *J. Food Qual.* **2018**, *2018*, 9090831. [CrossRef]
13. Yu, X.J.; Tu, S.S.; Li, Y.H.; Yagoub, A.E.A.; Ma, H.L.; Zhou, C.S. Effects of single-and tri-frequency ultrasound on self-assembly and characterizations of bionic dynamic rat stomach digestion of pepsin-soluble collagen from chicken leg skin. *Food Res. Int.* **2020**, *137*, 109710. [CrossRef]
14. Liu, H.; Xu, X.; Liu, J.Q.; Zhang, H.R.; Qi, L.W.; Zhang, C.H. Comparative assessment of bone collagen recovered from different livestock and poultry species: Microstructure, physicochemical characteristics and functional properties. *Int. J. Food Sci. Tech.* **2023**, *58*, 1597–1610. [CrossRef]
15. Thankachan, S.N.; Ilamaran, M.; Ayyadurai, N.; Shanmugam, G. Insights into the effect of artificial sweeteners on the structure, stability, and fibrillation of type I collagen. *Int. J. Biol. Macromol.* **2020**, *164*, 748–758. [CrossRef]
16. Ran, Y.Q.; Su, W.; Ma, L.; Wang, X.L.; Li, X.D. Insight into the effect of sulfonated chitosan on the structure, rheology and fibrillogenesis of collagen. *Int. J. Biol. Macromol.* **2021**, *166*, 1480–1490. [CrossRef]
17. Chandrapala, J.; Oliver, C.; Kentish, S.; Ashokkumar, M. Ultrasonics in food processing. *Ultrason. Sonochem.* **2012**, *19*, 975–983. [CrossRef]
18. Liao, W.; Xia, G.H.; Li, Y.C.; Shen, X.R.; Li, C. Comparison of characteristics and fibril-forming ability of skin collagen from barramundi (Lates calcarifer) and tilapia (Oreochromis niloticus). *Int. J. Biol. Macromol.* **2018**, *107*, 549–559. [CrossRef]
19. Li, Y.P.; Asadi, A.; Monroe, M.R.; Douglas, E.P. pH effects on collagen fibrillogenesis in vitro: Electrostatic interactions and phosphate binding. *Mater. Sci. Eng. C* **2009**, *29*, 1643–1649. [CrossRef]
20. Tang, L.L.; Chen, S.L.; Su, W.J.; Weng, W.Y.; Osako, K.; Tanaka, M. Physicochemical properties and film-forming ability of fish skin collagen extracted from different freshwater species. *Process Biochem.* **2015**, *50*, 148–155. [CrossRef]
21. Zhang, X.; Ookawa, M.; Tan, Y.K.; Ura, K.; Adachi, S.; Takagi, Y. Biochemical characterisation and assessment of fibril-forming ability of collagens extracted from Bester sturgeon Huso huso× Acipenser ruthenus. *Food Chem.* **2014**, *160*, 305–312. [CrossRef] [PubMed]
22. Meng, D.W.; Tanaka, H.; Kobayashi, T.; Hatayama, H.; Zhang, X.; Ura, K.; Yunoki, S.; Takagi, Y. The effect of alkaline pretreatment on the biochemical characteristics and fibril-forming abilities of types I and II collagen extracted from bester sturgeon by-products. *Int. J. Biol. Macromol.* **2019**, *131*, 572–580. [CrossRef] [PubMed]
23. Ferraro, V.; Gaillard-Martinie, B.; Sayd, T.; Chambon, C.; Anton, M.; Santé-Lhoutellier, V. Collagen type I from bovine bone. Effect of animal age, bone anatomy and drying methodology on extraction yield, self-assembly, thermal behaviour and electrokinetic potential. *Int. J. Biol. Macromol.* **2017**, *97*, 55–66. [CrossRef] [PubMed]
24. Yousefi, M.; Ariffin, F.; Huda, N. An alternative source of type I collagen based on by-product with higher thermal stability. *Food Hydrocolloid* **2017**, *63*, 372–382. [CrossRef]
25. Andonegi, M.; Las Heras, K.; Santos-Vizcaíno, E.; Igartua, M.; Hernandez, R.M.; de la Caba, K.; Guerrero, P. Structure-properties relationship of chitosan/collagen films with potential for biomedical applications. *Carbohyd. Polym.* **2020**, *237*, 116159. [CrossRef]
26. Valencia-Llano, C.H.; López-Tenorio, D.; Saavedra, M.; Zapata, P.A.; Grande-Tovar, C.D. Comparison of Two Bovine Commercial Xenografts in the Regeneration of Critical Cranial Defects. *Molecules* **2022**, *27*, 5745. [CrossRef]
27. Ion, A.; Andronescu, E.; Rădulescu, D.; Rădulescu, M.; Iordache, F.; Vasile, B.Ș.; Surdu, A.V.; Albu, M.G.; Maniu, H.; Chifiriuc, M.C.; et al. Biocompatible 3d matrix with antimicrobial properties. *Molecules* **2016**, *21*, 115. [CrossRef]
28. Du, X.; Li, H.J.; Nuerjiang, M.; Shi, S.; Kong, B.H.; Liu, Q.; Xia, X.F. Application of ultrasound treatment in chicken gizzards tenderization: Effects on muscle fiber and connective tissue. *Ultrason. Sonochem.* **2021**, *79*, 105786. [CrossRef] [PubMed]
29. Liu, D.S.; Zhou, P.; Li, T.C.; Regenstein, J.M. Comparison of acid-soluble collagens from the skins and scales of four carp species. *Food Hydrocolloid* **2014**, *41*, 290–297. [CrossRef]
30. Pal, G.K.; Suresh, P.V. Comparative assessment of physico-chemical characteristics and fibril formation capacity of thermostable carp scales collagen. *Mater. Sci. Eng. C* **2017**, *70*, 32–40. [CrossRef] [PubMed]
31. Dong, X.F.; Shen, P.; Yu, M.Q.; Yu, C.X.; Zhu, B.W.; Qi, H. (−)-Epigallocatechin gallate protected molecular structure of collagen fibers in sea cucumber Apostichopus japonicus body wall during thermal treatment. *LWT Food Sci. Technol.* **2020**, *123*, 109076. [CrossRef]

32. Adamiak, K.; Lewandowska, K.; Sionkowska, A. The infuence of salicin on rheological and film-forming properties of collagen. *Molecules* **2021**, *26*, 1661. [CrossRef]
33. Li, Y.S.; Yang, L.H.; Wu, S.J.; Chen, J.D.; Lin, H.W. Structural, functional, rheological, and biological properties of the swim bladder collagen extracted from grass carp (Ctenopharyngodon idella). *LWT Food Sci. Technol.* **2022**, *153*, 112518. [CrossRef]
34. Song, X.; Si, L.L.; Sun, X.; Zhu, X.; Li, Z.X.; Li, Y.Y.; Wang, Y.F.; Hou, H. Rheological properties, thermal stability and conformational changes of collagen from sea cucumber (Apostichopus japonicas). *Food Chem.* **2022**, *389*, 133033. [CrossRef] [PubMed]
35. Duan, L.; Li, J.H.; Li, C.H.; Li, G.Y. Effects of NaCl on the rheological behavior of collagen solution. *Korea-Aust. Rheol. J.* **2013**, *25*, 137–144. [CrossRef]
36. Lai, G.L.; Li, Y.; Li, G.Y. Effect of concentration and temperature on the rheological behavior of collagen solution. *Int. J. Biol. Macromol.* **2008**, *42*, 285–291. [CrossRef]
37. Pudło, A.; Juchniewicz, S.; Kopeć, W. Characteristics of Reconstituted Collagen Fibers from Chicken Keel Cartilage Depends on Salt Type for Removal of Proteoglycans. *Molecules* **2021**, *26*, 3538. [CrossRef]
38. Kang, D.C.; Wang, A.R.; Zhou, G.H.; Zhang, W.G.; Xu, S.M.; Guo, G.P. Power ultrasonic on mass transport of beef: Effects of ultrasound intensity and NaCl concentration. *Innov. Food Sci. Emerg.* **2016**, *35*, 36–44. [CrossRef]
39. Thuy, L.T.M.; Okazaki, E.; Osako, K. Isolation and characterization of acid-soluble collagen from the scales of marine fishes from Japan and Vietnam. *Food Chem.* **2014**, *149*, 264–270. [CrossRef]
40. Kang, D.C.; Zou, Y.H.; Cheng, Y.P.; Xing, L.J.; Zhou, G.H.; Zhang, W.G. Effects of power ultrasound on oxidation and structure of beef proteins during curing processing. *Ultrason. Sonochem.* **2016**, *33*, 47–53. [CrossRef]
41. Pezeshk, S.; Rezaei, M.; Abdollahi, M. Impact of ultrasound on extractability of native collagen from tuna by-product and its ultrastructure and physicochemical attributes. *Ultrason. Sonochem.* **2022**, *89*, 106129. [CrossRef] [PubMed]
42. Monago-Maraña, O.; Wold, J.P.; Rødbotten, R.; Dankel, K.R.; Afseth, N.K. Raman, near-infrared and fluorescence spectroscopy for determination of collagen content in ground meat and poultry by-products. *LWT-Food Sci. Technol.* **2021**, *140*, 110592. [CrossRef]
43. Qu, W.J.; Guo, T.T.; Zhang, X.X.; Jin, Y.T.; Wang, B.; Wahia, H.; Ma, H.L. Preparation of tuna skin collagen-chitosan composite film improved by sweep frequency pulsed ultrasound technology. *Ultrason. Sonochem.* **2022**, *82*, 105880. [CrossRef] [PubMed]

molecules

Article

Fabrication and Characterization of the Egg-White Protein Chitosan Double-Layer Emulsion

Qingmei Pan [1,2], Zhipeng Zhang [3], Ruifang Wu [1,2], Danjun Guo [1,2,*], Wei Xu [1,2,*], Hongxun Wang [1,2] and Yang Yi [1,2]

1 School of Food Science and Engineering, Wuhan Polytechnic University, Wuhan 430023, China
2 Key Laboratory for Deep Processing of Major Grain and Oil, Ministry of Education (Wuhan Polytechnic University), Wuhan 430023, China
3 BG (TianJin) Grain and Oil Industry Co., Ltd., Tianjin 300452, China
* Correspondence: missguodj@163.com (D.G.); xuwei1216@163.com (W.X.)

Abstract: Egg-white protein has an abundance of hydrophobic amino acids and could be a potential emulsifier after modification. Here, egg-white protein was modified via ultrasonic and transglutaminase treatments to destroy the globular structure. The egg-white protein gel particles (EWP-GPs) were prepared and then a novel highly stable EWP-chitosan double-layer emulsion was constructed. When ultrasonic treatment was applied at 240 W and TGase (20 U/g EWP) treatment, the EWP-GPs had a low particle size and good emulsification performance. The particle size of EWP-GPs was a minimum of 287 nm, and the polymer dispersity index (PDI) was 0.41. The three-phase contact angle ($\theta_{o/w}$) of EWP-GPs was 79.6° (lower than 90°), performing with good wettability. Based on these results, the EWP-chitosan double-layer emulsion was prepared through the EWP-GPs being treated with 240 W ultrasound, TGase, and chitosan in this study. When the double-layer emulsion had 0.6% (*v/v*) chitosan, the zeta potential of the double-layer emulsion was −1.1 mV and the double-layer emulsion had a small particle size (56.87 μm). The creaming index of double-layer emulsion at 0.6% (*v/v*) chitosan was 16.3% and the droplets were dispersed uniformly. According to the rheological results, the storage modulus (G′) was larger than the loss modulus (G″) in the whole frequency, indicating the formation of an elastic gel network structure in the emulsion. It is hoped to develop a novel food-grade stabilizer and a stable double-layer emulsion, providing new environment-friendly processing in hen egg products and delivery systems.

Keywords: egg-white protein; chitosan; particles; double-layer emulsion; characterization

Citation: Pan, Q.; Zhang, Z.; Wu, R.; Guo, D.; Xu, W.; Wang, H.; Yi, Y. Fabrication and Characterization of the Egg-White Protein Chitosan Double-Layer Emulsion. *Molecules* **2022**, *27*, 6036. https://doi.org/10.3390/molecules27186036

Academic Editor: Antonio-José Trujillo

Received: 18 August 2022
Accepted: 13 September 2022
Published: 16 September 2022

1. Introduction

Emulsion is a common material for delivering bioactive substances in food, pharmacy, and cosmetics [1], while single-layer emulsion is usually limited by its instability to environmental change [2]. Recently, double-layer emulsion has become a hot issue for avoiding instability constraints on environmental conditions [3,4]. The double-layer emulsion was a water-in-oil-in-water ($W_1/O/W_2$) system and it was generally prepared in two steps [5]. The first step was to prepare a W_1/O emulsion as the internal phase. Then, the W_1/O emulsion was dispersed in the external aqueous phase (W_2). It was reported that a $W_1/O/W_2$ emulsion should be stabilized with two kinds of emulsifiers (lipophilic to W_1/O and hydrophilic to O/W_2) to avoid the migration of the internal and external phases [6]. Additionally, a Pickering emulsion was stabilized through solid particles and exhibited high stability against coalescence [7]. Thus, Pickering emulsion is a potential internal phase to improve the stability of the double-layer emulsion [7–9]. Food-grade particles derived from proteins, polysaccharides, and lipids were reported to be used to stabilize the Pickering emulsions [10]. Among these particles, protein particles, as an excellent natural emulsifier, have high surface activity and good digestibility [11].

In recent years, many protein particles have been used to stabilize Pickering emulsions, such as whey protein isolate [12] and zein [13]. In fact, egg-white protein has superior gelation and is amphiphilic due to its abundant hydrophobic amino acids [14,15]. However, egg-white is a typical globular protein, and a large number of hydrophobic groups are embedded in the protein molecules and exhibit strong hydrophilic properties [16]. Protein modifications (chemical modification, physical modification, and enzymatic modification) could induce partial unfoldment and aggregation of globular proteins, so the hydrophobic groups could be exposed to the surface and show good emulsification. Among these modification methods, the ultrasonic and enzymatic modifications are safe, low-loss, and friendly to the environment [17]. In addition, ultrasonic modification combined with enzymolysis could improve the efficiency of protein modification and boost the biological activity of proteins [18]. Transglutaminase (TGase) is an acyltransferase, and it could boost the cross-linking and gelation properties of protein [19]. Studies found that ultrasound combined with TGase treatment was a superior processing technology which improved the solubility and emulsifying properties of protein particles, such as soy protein [20] and soybean-whey mixed protein [21]. However, protein particles may induce the aggregation of droplets and cause the instability of emulsions [22].

Chitosan, a natural polysaccharide, is the by-product of chitin deacetylation [23]. It has strong hydrophilic properties and good adhesion. Chitosan could change the viscosity of emulsions and provide repulsive electrostatics to prevent droplet aggregation [24]. A recent study found that whey protein isolate and chitosan hydrochloride were used to prepare double-layer emulsion via layer-by-layer self-assembly, which showed better stability than monolayer emulsions [25]. The purpose of this study was to obtain the EWP-GPs with good emulsification and construct a highly stable EWP-chitosan double-layer emulsion system. In this study, egg-white protein was first modified via ultrasonic and transglutaminase treatments, and EWP-GPs were prepared and characterized by particle size, fluorescent spectrometry, three-phase contact angle, and emulsification. Secondly, the EWP-chitosan double-layer emulsion was constructed and investigated through the particle size, zeta potential, rheology, and microstructures. This study is expected to expand the application of egg-white as a stabilizer and obtain a food-grade double-layer emulsion to deliver bioactive substances.

2. Results and Discussion

2.1. Preparation and Properties of Egg-White Protein Gel Particles

2.1.1. Particle Size and Wettability

The polymer dispersity index (PDI) was used to describe the molecular weight distribution of polymers [26]. As shown in Figure 1a, the particle size of EWP-GPs without ultrasonic treatment was 439.17 nm. When the ultrasonic power was 360 W, the particle size of EWP-GPs significantly decreased to the smallest size at 412 nm ($p < 0.05$), and the PDI was at a minimum of 0.32. The PDI of EWP-GPs with 360 W ultrasonic treatment was significantly lower than the EWP-GPs without ultrasonic treatment ($p < 0.05$), indicating that 360 W ultrasonic treatment could improve the dispersion and stability of EWP-GPs. This phenomenon might be attributable to the mechanical shear produced by the ultrasonic treatment [27]. A study prepared a flaxseed oil-in-water emulsion and found a similar phenomenon: with the improvement of the ultrasonic power (generation at 20–24 kHz), the mechanical shear was strengthened, and the particle size and PDI were decreased [28]. Overall, a 360 W ultrasonic treatment could provide EWP-GPs with good dispersibility.

Figure 1b showed the particle size of the EWP-GPs prepared under TGase treatment combined with differing ultrasonic power (0–600 W). The size of EWP-GPs prepared with TGase but without ultrasonic treatment was 307.83 nm, which was much smaller than the EWP-GPs without treatment (439.17 nm). This was because transglutaminase improved the jelly strength of egg-white protein and produced smaller particles in preparation [29]. With 240 W ultrasonic and TGase treatment, the EWP-GPs reached the smallest particle size (287 nm), and the PDI was 0.41. This phenomenon suggested that ultrasonic and

TGase treatment could decrease the particle size of EWP-GPs and improve the stability of particles. As Figure 1a,b show, the PDI value of EWP-GPs with ultrasonic-TGase treatment was higher than that of EWP-GPs with ultrasonic treatment. This might be due to the jelly strength of EWP-GPs being improved by TGase treatment, and uneven particles were generated in preparation.

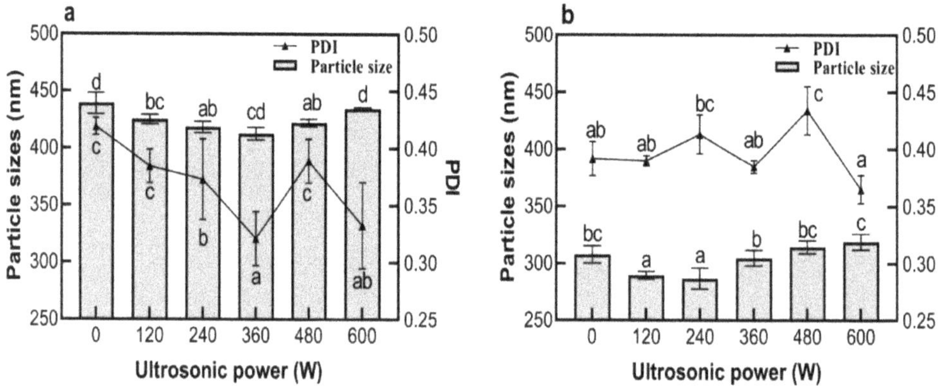

Figure 1. Effect of ultrasonic power on the particle size of EWP-GPs (**a**) and EWP-GPs with TGase treatment (**b**) of EWP-GPs. (Values with different letters are significant at $p < 0.05$).

As a Pickering particle, the particle should be insoluble but partially wetted by the continuous and dispersed phases. Partial wettability of particles contributed to sufficient interfacial adsorption efficiency. A three-phase contact angle of particles was the angle formed at the three-phase boundary where the solid particles, the continuous phase, and the dispersed phase intersect. The wettability of EWP-GPs could be evaluated by the three-phase contact angle ($\theta_{o/w}$), which could evaluate the adsorption efficiency of the emulsion. When the $\theta_{o/w}$ was less than 90°, the particles showed a good hydrophilicity. When the $\theta_{o/w}$ was more than 90°, the particles were inclined to show a good hydrophobicity. When the $\theta_{o/w}$ was 90°, the particles had the strongest absorption and could prevent the aggregation of droplets [2]. As shown in Figure 2, the $\theta_{o/w}$ of EWP-GPs without any treatment was 52°: this indicated that the hydrophobicity of the particle was weak. The $\theta_{o/w}$ of EWP-GPs with 600 W ultrasonic treatment reached a peak (67.5°). These results indicated that the ultrasonic treatment could improve the hydrophobicity of EWP-GPs and that the EWP-GPs were partially wetted. It could be inferred that ultrasonic treatment caused the exposure of the protein chains and hydrophobic groups. A study suggested that ultrasonic power could destroy the non-covalent interaction of the whey proteins' particles and enhance the hydrophobicity of particles [30]. The $\theta_{o/w}$ of EWP-GPs with 240 W ultrasonic and TGase treatment was the largest (79.6°). This result showed that ultrasonic and TGase treatment could improve the hydrophile and lipophile of EWP-GPs. Transglutaminase could catalyze the formation of covalent bonds between lysine and glutamic acid in protein amino acids [31]. In addition, the ultrasonic treatment could induce the protein to unfold and the hydrophobic groups' exposure, resulting in facilitating the cross-link between an enzyme and a protein [32]. A previous study also found that ultrasound pretreatment promoted the enzymatic hydrolysis of soy protein isolates and then improved their surface hydrophobicity [31]. Overall, when EWP-GPs were treated with 240 W ultrasonic power combined with TGase, the particles performed a superior amphipathicity to prepare O/W emulsion.

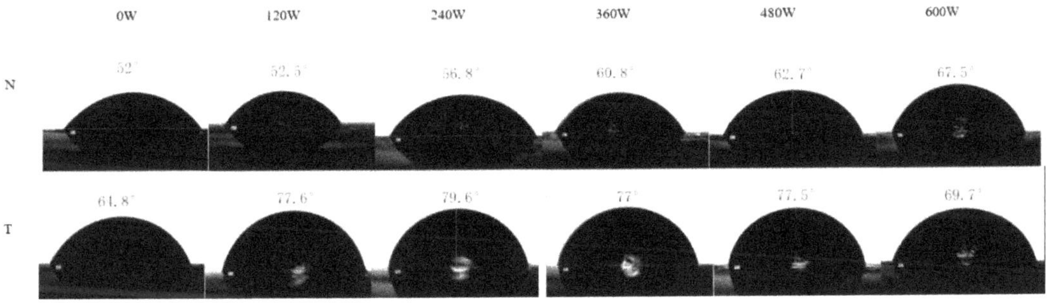

Figure 2. The effect of different treatments on the three-phase contact angle. (The EWP-GPs were treated with ultrasound from 0 W to 600 W, respectively. The "N" represents EWP-GPs without TGase treatment, and the "T" represents EWP-GPs with TGase treatment).

2.1.2. Zeta Potential of EWP-GPs and Chitosan

The double-layer emulsion had higher stability and protein–polysaccharides (such as a chitosan compound which was a hot issue to prepare double-layer emulsion in recent years). Egg-white protein contains a variety of proteins, and the isoelectric point is between 4 and 5 [14]. Therefore, when the pH value was less than 5, the dispersion surface of EWP-GPs was positively charged. Chitosan is a cationic polysaccharide with positive charges in a wide range of pH values. In addition, the aqueous solution of chitosan is positively charged at pH 6.0 [33]. As Figure 3a shows, when the pH was 6.0, the zeta potential of EWP-GPs and chitosan solution were −10.18 mV and 19.33 mV, respectively. The zeta potential of EWP-GPs and the chitosan solution were opposite, and the values were both higher than 10 mV. It was suitable to prepare stable emulsions. A previous study found that opposite charges between zein and gum Arabic caused gum Arabic to absorb on the surface of zein particles [13]. Through mechanical stirring, the cationic polyelectrolyte chitosan was adsorbed to the emulsion interface by electrostatic interaction to form an ultra-thin polymer interfacial film [23]. In addition, the electrostatic attraction between EWP-GPs and chitosan could be used to prepare the double-layer emulsion. Therefore, we could adjust the pH value to 6 to prepare the double-layer emulsion.

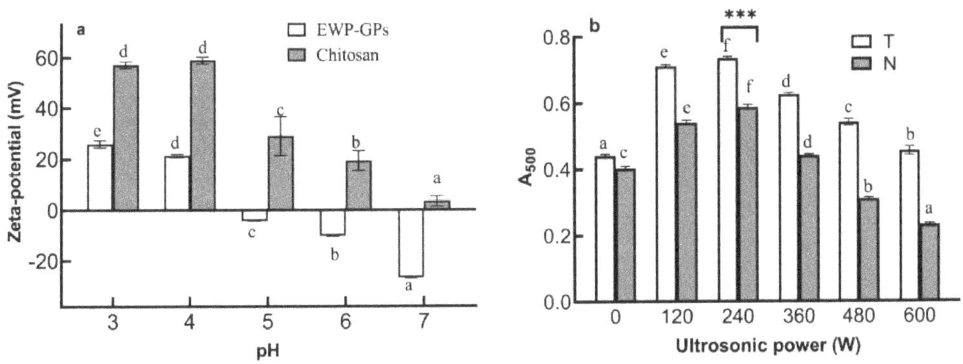

Figure 3. The effect of pH on the potential of EWP-GPs and chitosan (**a**). The emulsification of EWP-GPs (**b**). (Values with different letters are significant at $p < 0.05$, and *** represents $p < 0.001$).

2.1.3. Emulsification of EWP-GPs

Emulsification indicates the ability of a protein to adsorb at the oil–water interface. As Figure 3b showed, when the EWP-GPs were treated with 240 W ultrasonic and TGase, the absorbance of EWP-GPs reached the maximum (0.74). In addition, the TGase significantly increased the emulsification of EWP-GPs with 240 W ultrasonic-TGase treatment ($p < 0.001$) compared with the EWP-GPs with 240 W ultrasonic treatment. It revealed that ultrasonic and TGase treatment could improve the emulsification of EWP-GPs. Ultrasonic and enzymatic treatment could increase the reaction sites and improve the emulsification of particles [34,35]. The emulsification of EWP-GPs was bound up with the surface hydrophobicity of EWP-GPs. A study found a similar phenomenon in Cyperus esculentus seed protein [36]. The EWP-GPs with 240 W ultrasonic-TGase treatment had high surface hydrophobicity and emulsification, and the particles could be used to prepare emulsions.

2.1.4. Intermolecular Force

As Figure 4a shows, the transmittance of the particles in the control group presented irregular changes with the increase in ultrasonic power. These results were not consistent with the changes of particle size. The reason might be that the ultrasonic power promoted the aggregation of EWP-GPs. As Figure 4b shows, the transmittance of the particles in the control group with 240 W ultrasonic-TGase treatment reached the highest value (7.5%): this was consistent with the result of particle size. This also suggested that the smaller particles had higher transmittance. As Figure 4a,b show, the transmittances of EWP-GP dispersions were significantly increased with the addition of SDS, Urea, or DTT ($p < 0.05$). It was found that SDS, Urea, and DTT could destroy the hydrophobic interaction, hydrogen bond, and disulfide bond inside the particles, respectively [37]. Therefore, the hydrophobic interaction, hydrogen bond, and disulfide bond were the intramolecular interactive forces maintaining the internal structure of the EWP-GPs. When the EWP-GPs were untreated with ultrasound and TGase, the transmittances in Urea, SDS, and DTT groups increased by 10%, 5%, and 1% of particles compared with the transmittances of the control group, respectively. The results showed that the hydrogen bond was the main force to maintain the stability of EWP-GPs.

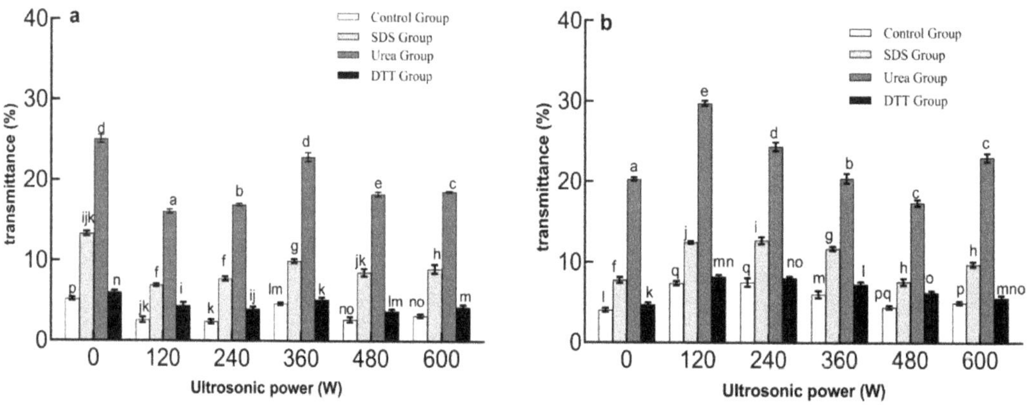

Figure 4. The transmittance of ultrasonic treatment EWP-GPs (**a**) and ultrasonic-TGase treatment EWP-GPs (**b**). (Values with different letters are significant at $p < 0.05$).

In conclusion, when the EWP-GPs were treated with 240 W ultrasound and TGase, the particles had small particles and good emulsification, which was suitable to prepare highly stable double-layer emulsions. In addition, when the pH value was 6, the EWP-GPs and chitosan were suitable to prepare the double-layer emulsion.

2.2. *Characteristics of Double-Layer Emulsions*

2.2.1. Zeta Potential and Particle Size of the Double-Layer Emulsion

Figure 5a showed that when the concentration of chitosan was 0.6%, the positive and negative charges on the surface of the emulsion tended to balance, and the zeta potential of the double-layer emulsion was −1.1 mV. When the pH was 6, the potential of EWP-GPs was −10.18 mV and the potential of chitosan solution was 19.33 mV. With the increase in the concentration of chitosan, the zeta potential of the double-layer emulsion changed from negative to positive. Therefore, it could be inferred that the chitosan absorbed on the surface of EWP-GPs in the emulsion. A study found similar results: that the chitosan molecules could be adsorbed on the surface of WPI gel particles and the zeta potential changed with the increase in chitosan content [38]. The electrostatic interaction could not maintain the stability of the emulsion when the concentration of chitosan exceeded 0.6%.

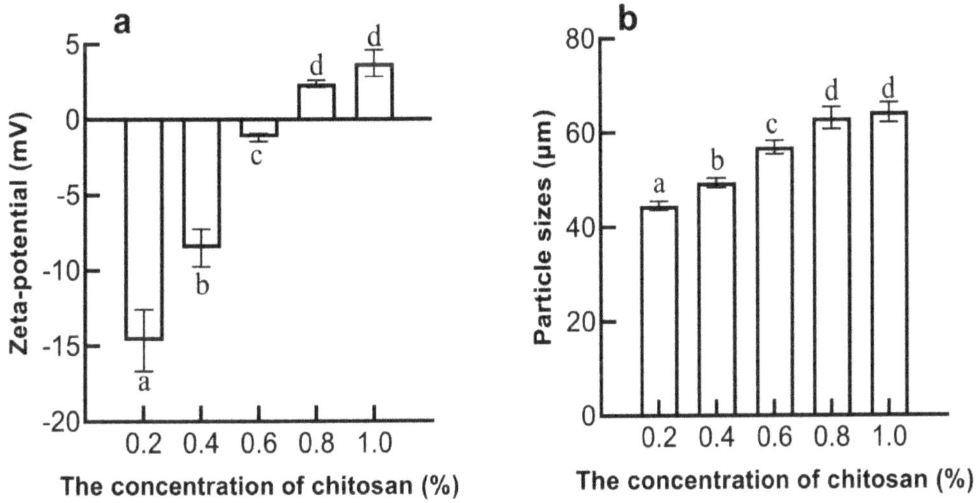

Figure 5. The effect of chitosan concentration on the potential of the double-layer emulsion (**a**). The effect of chitosan concentration on the particle size of the double-layer emulsion (**b**). (Values with different letters are significant at $p < 0.05$).

As Figure 5b shows, when the chitosan concentration was 0.6%, the particle size of the double-layer emulsion was 56.87 μm. With the addition of chitosan, the particle size of the double-layer emulsion increased, indicating that chitosan could be adsorbed on the surface of EWP-GP droplets to form a thicker interface layer [39]. When the chitosan concentration was greater than 0.8%, the particle size of the double-layer emulsion had no significant change ($p > 0.05$). This might be due to the saturation of chitosan adsorption on the surface of the emulsion when the concentration of chitosan exceeded 0.8%. When the concentration of chitosan was less than 0.6%, chitosan molecules were adsorbed between the droplets due to electrostatic interaction. A study also found that electrostatic interaction between egg-white protein and κ-Carrageenan led to an increase in bridging flocculation and particle size [40].

2.2.2. Microstructures and Stability of Double-Layer Emulsion

Figure 6 showed the microscopic image and macro view of double-layer emulsions with different chitosan concentrations. When the concentration of chitosan was 0.2% and 0.6%, the droplets dispersed uniformly. When the concentration of chitosan exceeded 0.6%, most of the drops gathered together. This might be due to the charge on the surface of the emulsion droplets being close to 0, and the electrostatic repulsion between the emulsion droplets decreased.

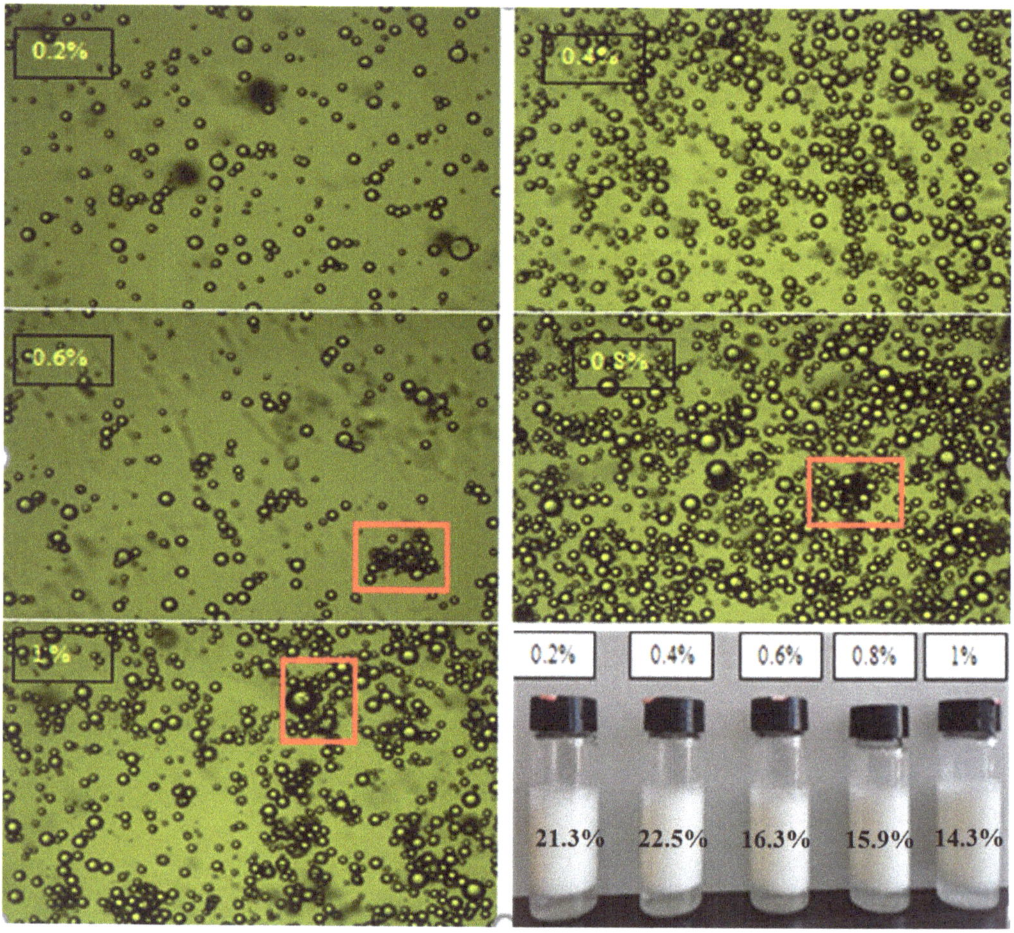

Figure 6. Microscopic image and macro view of double-layer emulsion. The graphs were collected at 100× magnification. (The red boxes point out that the emulsion drops gathered together. The creaming indexes of double-layer emulsions are labeled in the picture).

As was shown in Figure 6, the creaming index (CI%) of the double-layer emulsion decreased as the chitosan concentration increased from 0.4% to 1.0%, exhibiting better creaming stability. With the increase in chitosan concentration, chitosan molecules could form a film on the surface of the double-layer emulsion to prevent emulsion aggregation and improve the stability of the emulsion [41].

2.2.3. Rheology Analysis

The apparent shear viscosity of three emulsions decreased with an increase in shear rate (Figure 7a), indicating that they were non-Newtonian shear-thinning fluids [42]. At a low shear rate (less than 10/s), the apparent viscosity of the double-layer emulsion increased with the increase in chitosan concentration. This might be due to the internal resistance of the emulsion during flow, which hindered the free movement of the medium and led to a higher apparent viscosity [43]. With the increase in shear rate (more than 10/s), all of the droplets rearranged from disorder to order, and the apparent viscosity decreased. This might have occurred because any flocs or clumps in the emulsions had broken down at high shear rates. The shear viscosity hardly changed in the shear rate ranging from 10^2 to 10^3 s^{-1}, indicating that the emulsions exhibited ideal fluid characteristics [44].

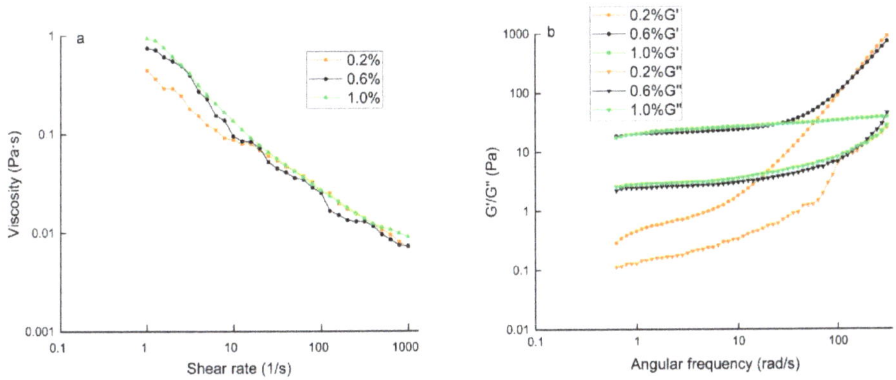

Figure 7. The viscosity of double-layer emulsions with various chitosan concentrations (**a**). Frequency dependence of the storage (G′) and loss (G″) moduli for double-layer emulsions with various chitosan concentrations (**b**).

Storage modulus (G′) represents the elastic behavior of the emulsion network structure. The loss modulus (G″) represents the viscous behavior of the emulsion network structure. Figure 7b shows that G′ of each double-layer emulsion was much higher than that of G″ over the whole frequency range, suggesting that the gel network structure dominated by elasticity was formed in the emulsion [45,46]. When the chitosan concentration was 0.6%, the G′ of the double-layer emulsion was the highest. This phenomenon indicated that the interaction between droplets improved, and the ability to resist the deformation of the double-layer emulsion was strong. Meanwhile, the addition of chitosan molecules also improved the viscosity and contributed to the strength of the gel structure.

3. Materials and Methods

3.1. Materials and Reagents

The hen eggs were purchased from the Beijing CP Egg Industry Co., Ltd. (Beijing, China). Soybean oil (Yihai Kerry Arawana Holdings Co., Ltd.) was purchased from a local market (Wuhan, China). Glutamine transaminase (TGase, 100 U/g), chitosan, a Coomassie Brilliant Blue kit, Sodium Dodecyl Sulfate (SDS), Dithiothreitol (DTT), and ANS were purchased from Yuanye Bio-Technology Co., Ltd. (Shanghai, China).

3.2. The Preparation of Egg-White Protein Gel Particles

The egg-white protein gel particles (EWP-GPs) were prepared according to a previous method [47]. The egg-white was separated from fresh hen eggs and filtered with gauze to remove insoluble substances after stirring (1 h at 25 °C). The protein concentration of the solution was 10%, measured via Coomassie Brilliant Blue [48]. The EWP solution (30 mL) was ultrasonicated for 10 min at 120 W. The pH of the solution was adjusted to 7.0 with NaOH solution (1 mol/L) and HCl solution (1 mol/L). The EWP solution was stirred at 25 °C for 2 h and then reacted with TGase (20 U/g EWP) at 45 °C in a water bath for 2 h. The mixture of EWP and chitosan was heated at 90 °C for 40 min and then cooled in an ice bath immediately and maintained at 4 °C for 24 h. The gel particles were obtained, crushed, and diluted with deionized water. The gel particles were pre-homogenized for 2 min with a high-speed disperser (XHF-DY, Scientz, Ningbo, China) at 10,000 rpm. Finally, the EWP-GPs were obtained through homogenizing three times at 20,000 psi with a high-pressure micro-jet homogenizer (APV1000, APV Co., Crawley, UK).

3.3. *Characterization of EWP-GPs*

3.3.1. Particle Size and Zeta Potential

In accordance with a previous method [26] and slightly modified, the particle size, PDI, and zeta potential of EWP-GPs and emulsions were determined via the particle electrophoresis instrument (Zetasizer Nano-ZS90, Malvern Instruments Ltd., Worcestershire, UK). The concentration of the EWP-GP solutions was diluted to 0.1% with deionized water to avoid multiple light scattering. The chitosan solution was diluted to 0.1% with deionized water and the zeta potential of the chitosan solution was determined through the particle electrophoresis instrument. Each sample was measured for three cycles and scanned 12 times per cycle.

3.3.2. Wettability Measurement

In accordance with a previous method [49] with slight modifications, the wettability of EWP-GPs was evaluated through the three-phase contact angle ($\theta_{o/w}$) which was determined by using a VCA Optima system (AST Products Inc., Billerica, MA, USA). The fresh EWP-GPs were diluted to 0.1% and dropped onto the mica sheet. After drying at 25 °C for 16 h, the water droplets (2 μL) were added to the composite particles to determine the $\theta_{o/w}$. The $\theta_{o/w}$ was fitted to the Laplace-Young equation [50]. Each sample was repeated at least three times, and the average value was obtained.

3.3.3. Emulsification

In accordance with a previous method [34] with slight modifications, the mixture of 30 mL EWP-GP solution (2%) and 10 mL soybean oil was treated via a high-speed homogenizer at 12,500 rpm for 2 min. The emulsion (50 μL) was removed from the bottom of the beaker, and then diluted 100 times with 0.1% (*w/v*) SDS. The absorbance of the emulsion at 500 nm was represented by the emulsification of the EWP-GPs.

3.3.4. Intermolecular Force

The intermolecular force of EWP-GPs was determined by measuring the absorbance of the solution at the wavelength of 600 nm at 25 °C with a UV/VIS spectrophotometer (UV2000, UNICO (SHANGHAI) INSTRUMENT CO., LTD., China) via a previous means [51] with modifications. The pattern of intermolecular forces involved in forming and maintaining the EWP-GP structure was investigated by determining the absorbance of dispersion with different protein denaturants (6 mol/L urea, 0.5% (*w/w*) SDS, and 30 mmol/L DTT). The control group was not treated with any protein denaturant (urea, SDS, or DTT solutions). After 10 min of reaction, the EWP-GP solution was diluted to 0.1% (*w/w*) to remain in the linear region of absorbance. Each sample was measured at least three times in parallel, and the average value was taken. The transmittance (T) was calculated by the following formula:

$$A = lg\frac{1}{T} \quad (1)$$

where A represented the absorbance of the solution and T represented the transmittance of the solution.

3.4. *Egg-White Protein Chitosan Double-Layer Emulsion Preparation*

In accordance with the previous study with slight modifications, the double-layer emulsion was prepared in two steps [7]. Firstly, the egg-white protein pickering emulsion (W_1/O emulsion) was prepared with EWP-GP dispersion (4 wt.% EWP-GPs) and 40% (*v/v*) soybean oil, and the mixture was dispersed via a high-speed homogenizer at 15000 rpm for 2 min. Secondly, 0.2%, 0.4%, 0.6%, 0.8% and 1.0% (*v/v*) chitosan solutions (W_2) were added to the Pickering emulsion, respectively. The mixture was dispersed via a high-speed homogenizer at 15,000 rpm for 2 min, and then egg-white protein-chitosan double-layer emulsion was obtained.

3.5. Double-Layer Emulsion Characterization

3.5.1. Particle Size and Zeta Potential

The method to detect the particle size and zeta potential of the double-layer emulsion was the same as the method in Section 3.3.1. In addition, the emulsions were diluted to 0.1% with deionized water to avoid multiple light scattering. Each sample was measured for three cycles and scanned 12 times per cycle.

3.5.2. Emulsion Microstructure

In accordance with the reported means [52] with modifications, the microstructure of the emulsion was observed through an optical microscope (SG–51, Shanghai optical instrument factory, Shanghai, China). The emulsion (20 μL) was placed on a microscope slide and covered by a cover slip. After balancing for 2 min, the microphotographs were collected. The graphs were collected at 100× magnification.

3.5.3. Creaming Index (*CI*)

The creaming index was measured via a reported method [23] with modifications. Fresh emulsion (8.0 mL) was added to the test vials and sealed with lids immediately after preparation. The height of the serum phase at the bottom (*Hs*) and the total height of emulsions (*Ht*) were recorded at fixed intervals. The experiment was carried out at 25 °C and the percentage of *CI* (%) was calculated using the following equation:

$$CI(\%) = \frac{Hs}{Ht} \times 100 \quad (2)$$

where *Hs* represented the height of the serum phase and *Ht* represented the total height of emulsions.

3.5.4. Rheology Measurement

The rheology of the double-layer emulsions was performed with a Discovery DHR-2 Rheometer (TA Instruments, New Castle, DE, USA) by a previous means [46] with modifications. The samples were placed between two parallel plates with a diameter of 40 mm at 25 °C. The static shear rheological properties of the emulsions were measured at a shear rate from 0.1 to 100 rad/s. The storage modulus (G') and loss modulus (G'') of the emulsions were recorded as a function of frequency from 0.1 to 16 Hz.

3.6. Statistical Analysis

Data were presented as mean ± standard deviation (SD) of three independent experiments. The statistical analysis was assessed using one-way analysis of ANOVA with Duncan's multiple range test. p-values < 0.05 were considered as statistically significant.

4. Conclusions

In this study, the EWP-GPs and EWP–chitosan double-layer emulsion were successfully prepared. The results showed that ultrasonic treatment combined with TGase treatment could significantly improve the surface hydrophobicity and stability of EWP-GPs. The optimal EWP-GPs had a smaller particle size, higher dispersity, better emulsification, and good wettability. The intermolecular force results showed that the hydrogen bond was the main force to maintain the stability of EWP-GPs. Based on these results, the EWP-chitosan double-layer emulsion was prepared and characterized. The double-layer emulsion had a small particle size and the droplets were dispersed uniformly, suggesting high stability. The rheological properties of the double-layer emulsion indicated that the gel network structure dominated by elasticity was formed in the emulsion. The prepared EWP-chitosan double-layer emulsion performed with high stability and good creaming stability. In a further study, we could study the delivery functions of the EWP-chitosan emulsion. It could provide a novel food-grade delivery system for bioactive substances, such as β-carotene and polyphenol.

Author Contributions: Q.P.: data curation, visualization, formal analysis, writing—original draft. Z.Z.: conceptualization, methodology, formal analysis, data curation. R.W.: formal analysis, data curation. D.G.: writing—review and editing, funding acquisition. W.X.: conceptualization, project administration, writing—review and editing. H.W.: writing—review and editing. Y.Y.: writing—review and editing. All authors have read and agreed to the published version of the manuscript.

Funding: This work was financially supported by the Scientific Research Project of the Education Department of Hubei Province (No. B2021124).

Institutional Review Board Statement: Not applicable.

Informed Consent Statement: Not applicable.

Data Availability Statement: Not applicable.

Conflicts of Interest: The authors declare no conflict of interest.

Abbreviations

EWP-GPs	Egg-white protein gel particles
TGase	Transglutaminase
PDI	Polymer dispersity index
CI	Creaming index

References

1. Ye, F.; Miao, M.; Cui, S.W.; Jiang, B.; Jin, Z.Y.; Li, X.F. Characterisations of oil-in-water Pickering emulsion stabilized hydrophobic phytoglycogen nanoparticles. *Food Hydrocoll.* **2018**, *76*, 78–87. [CrossRef]
2. Feng, Y.; Lee, Y. Surface modification of zein colloidal particles with sodium caseinate to stabilize oil-in-water pickering emulsion. *Food Hydrocoll.* **2016**, *56*, 292–302. [CrossRef]
3. Fioramonti, S.A.; Martinez, M.J.; Pilosof, A.M.R.; Rubiolo, A.C.; Santiago, L.G. Multilayer emulsions as a strategy for linseed oil microencapsulation: Effect of pH and alginate concentration. *Food Hydrocoll.* **2015**, *43*, 8–17. [CrossRef]
4. Sagalowicz, L.; Leser, M.E. Delivery systems for liquid food products. *Curr. Opin. Colloid Interface Sci.* **2010**, *15*, 61–72. [CrossRef]
5. Silva, W.; Torres-Gatica, M.F.; Oyarzun-Ampuero, F.; Silva-Weiss, A.; Robert, P.; Cofrades, S.; Gimenez, B. Double emulsions as potential fat replacers with gallic acid and quercetin nanoemulsions in the aqueous phases. *Food Chem.* **2018**, *253*, 71–78. [CrossRef]
6. Klojdova, I.; Troshchynska, Y.; Stetina, J. Influence of carrageenan on the preparation and stability of w/o/w double milk emulsions. *Int. Dairy J.* **2018**, *87*, 54–59. [CrossRef]
7. Choi, M.J.; Choi, D.; Lee, J.; Jo, Y.J. Encapsulation of a bioactive peptide in a formulation of W-1/O/W-2-type double emulsions: Formation and stability. *Food Struct.* **2020**, *25*, 100145. [CrossRef]
8. He, D.Y.; Wu, Z.Z.; Cui, B.; Jin, Z.Y. A novel SERS-based aptasensor for ultrasensitive sensing of microcystin-LR. *Food Chem.* **2019**, *278*, 197–202. [CrossRef]
9. Li, X.M.; Zhu, J.; Pan, Y.; Meng, R.; Zhang, B.; Chen, H.Q. Fabrication and characterization of pickering emulsions stabilized by octenyl succinic anhydride-modified gliadin nanoparticle. *Food Hydrocoll.* **2019**, *90*, 19–27. [CrossRef]
10. Mwangi, W.W.; Lim, H.P.; Low, L.E.; Tey, B.T.; Chan, E.S. Food-grade Pickering emulsions for encapsulation and delivery of bioactives. *Trends Food Sci. Technol.* **2020**, *100*, 320–332. [CrossRef]
11. Yan, X.J.; Ma, C.C.; Cui, F.Z.; McClements, D.J.; Liu, X.B.; Liu, F.G. Protein-stabilized Pickering emulsions: Formation, stability, properties, and applications in foods. *Trends Food Sci. Technol.* **2020**, *103*, 293–303. [CrossRef]
12. Wang, X.; Li, X.; Xu, D.X.; Liu, G.R.; Xiao, J.S.; Cao, Y.P.; Sun, B.G. Influence of unadsorbed emulsifiers on the rheological properties and structure of heteroaggregate of whey protein isolate (WPI) coated droplets and flaxseed gum (FG) coated droplets. *Food Hydrocoll.* **2018**, *80*, 42–52. [CrossRef]
13. Dai, L.; Sun, C.X.; Wei, Y.; Mao, L.K.; Gao, Y.X. Characterization of Pickering emulsion gels stabilized by zein/gum arabic complex colloidal nanoparticles. *Food Hydrocoll.* **2018**, *74*, 239–248. [CrossRef]
14. Xiao, N.H.; Huang, X.; He, W.; Yao, Y.; Wu, N.; Xu, M.S.; Du, H.Y.; Zhao, Y.; Tu, Y.G. A review on recent advances of egg byproducts: Preparation, functional properties, biological activities and food applications. *Food Res. Int.* **2021**, *147*, 110563. [CrossRef] [PubMed]
15. Tang, C.H. Globular proteins as soft particles for stabilizing emulsions: Concepts and strategies. *Food Hydrocoll.* **2020**, *103*, 105664. [CrossRef]
16. Yu, Y.L.; Guan, Y.; Wen, H.D.; Zhang, Y.; Liu, J.B.; Zhang, T. Mild heating assisted alkaline pH shifting modify the egg white protein: The mechanism and the enhancement of emulsifying properties. *LWT Food Sci. Technol.* **2021**, *151*, 112094. [CrossRef]
17. Lv, X.H.; Huang, X.; Ma, B.; Chen, Y.; Batool, Z.; Fu, X.; Jin, Y.G. Modification methods and applications of egg protein gel properties: A review. *Compr. Rev. Food Sci. Food Saf.* **2022**, *21*, 2233–2252. [CrossRef]

18. Umego, E.C.; He, R.H.; Ren, W.B.; Xu, H.N.; Ma, H.L. Ultrasonic-assisted enzymolysis: Principle and applications. *Process Biochem.* **2021**, *100*, 59–68. [CrossRef]
19. Kuraishi, C.; Yamazaki, K.; Susa, Y. Transglutaminase: Its utilization in the food industry. *Food Rev. Int.* **2001**, *17*, 221–246. [CrossRef]
20. Geng, M.J.; Wang, Z.K.; Qin, L.; Taha, A.; Du, L.X.; Xu, X.Y.; Pan, S.Y.; Hu, H. Effect of ultrasound and coagulant types on properties of beta-carotene bulk emulsion gels stabilized by soy protein. *Food Hydrocoll.* **2022**, *123*, 107146. [CrossRef]
21. Cui, Q.; Wang, G.R.; Gao, D.; Wang, L.; Zhang, A.Q.; Wang, X.B.; Xu, N.; Jiang, L.Z. Improving the gel properties of transgenic microbial transglutaminase cross-linked soybean-whey mixed protein by ultrasonic pretreatment. *Process Biochem.* **2020**, *91*, 104–112. [CrossRef]
22. Ashaolu, T.J.; Zhao, G.H. Fabricating a Pickering Stabilizer from Okara Dietary Fibre Particulates by Conjugating with Soy Protein Isolate via Maillard Reaction. *Foods* **2020**, *9*, 143. [CrossRef] [PubMed]
23. Fu, J.J.; Zhu, Y.L.; Cheng, F.S.; Zhang, S.L.; Xiu, T.T.; Hu, Y.; Yang, S. A composite chitosan derivative nanoparticle to stabilize a W-1/O/W-2 emulsion: Preparation and characterization. *Carbohydr. Polym.* **2021**, *256*, 117533. [CrossRef] [PubMed]
24. Ribeiro, E.F.; Morell, P.; Nicoletti, V.R.; Quiles, A.; Hernando, I. Protein- and polysaccharide-based particles used for Pickering emulsion stabilisation. *Food Hydrocoll.* **2021**, *119*, 106839. [CrossRef]
25. Li, S.Q.; Sun, J.L.; Yan, J.; Zhang, S.R.; Shi, C.; McClements, D.J.; Liu, X.B.; Liu, F.G. Development of antibacterial nanoemulsions incorporating thyme oil: Layer-by-layer self-assembly of whey protein isolate and chitosan hydrochloride. *Food Chem.* **2021**, *339*, 128016. [CrossRef] [PubMed]
26. Dai, L.; Sun, C.X.; Wang, D.; Gao, Y.X. The Interaction between Zein and Lecithin in Ethanol-Water Solution and Characterization of Zein-Lecithin Composite Colloidal Nanoparticles. *PLoS ONE* **2016**, *11*, e0167172. [CrossRef]
27. Cui, Q.; Wang, X.B.; Wang, G.; Li, R.; Wang, X.D.; Chen, S.; Liu, J.N.; Jiang, L.Z. Effects of ultrasonic treatment on the gel properties of microbial transglutaminase crosslinked soy, whey and soy-whey proteins. *Food Sci. Biotechnol.* **2019**, *28*, 1455–1464. [CrossRef]
28. Kentish, S.; Wooster, T.J.; Ashokkumar, A.; Balachandran, S.; Mawson, R.; Simons, L. The use of ultrasonics for nanoemulsion preparation. *Innov. Food Sci. Emerg. Technol.* **2008**, *9*, 170–175. [CrossRef]
29. Er, B.; Sert, D.; Mercan, E. Production of skim milk powder by spray-drying from transglutaminase treated milk concentrates: Effects on physicochemical, powder flow, thermal and microstructural characteristics. *Int. Dairy J.* **2019**, *99*, 104544. [CrossRef]
30. Desrumaux, A.; Marcand, J. Formation of sunflower oil emulsions stabilized by whey proteins with high-pressure homogenization (up to 350 MPa): Effect of pressure on emulsion characteristics. *Int. J. Food Sci. Technol.* **2002**, *37*, 263–269. [CrossRef]
31. Chen, L.; Chen, J.S.; Ren, J.Y.; Zhao, M.M. Effects of Ultrasound Pretreatment on the Enzymatic Hydrolysis of Soy Protein Isolates and on the Emulsifying Properties of Hydrolysates. *J. Agric. Food Chem.* **2011**, *59*, 2600–2609. [CrossRef] [PubMed]
32. Penas, E.; Prestamo, G.; Gomez, R. High pressure and the enzymatic hydrolysis of soybean whey proteins. *Food Chem.* **2004**, *85*, 641–648. [CrossRef]
33. Wang, X.Y.; Heuzey, M.C. Chitosan-Based Conventional and Pickering Emulsions with Long-Term Stability. *Langmuir* **2016**, *32*, 929–936. [CrossRef]
34. Banach, J.C.; Lin, Z.; Lamsal, B.P. Enzymatic modification of milk protein concentrate and characterization of resulting functional properties. *LWT Food Sci. Technol.* **2013**, *54*, 397–403. [CrossRef]
35. Liang, Y.C.; Patel, H.; Matia-Merino, L.; Ye, A.Q.; Golding, M. Effect of pre- and post-heat treatments on the physicochemical, microstructural and rheological properties of milk protein concentrate-stabilised oil-in-water emulsions. *Int. Dairy J.* **2013**, *32*, 184–191. [CrossRef]
36. Cui, Q.; Wang, L.; Wang, G.R.; Zhang, A.Q.; Wang, X.B.; Jiang, L.Z. Ultrasonication effects on physicochemical and emulsifying properties of Cyperus esculentus seed (tiger nut) proteins. *LWT Food Sci. Technol.* **2021**, *142*, 110979. [CrossRef]
37. Schmitt, C.; Moitzi, C.; Bovay, C.; Rouvet, M.; Bovetto, L.; Donato, L.; Leser, M.E.; Schurtenberger, P.; Stradner, A. Internal structure and colloidal behaviour of covalent whey protein microgels obtained by heat treatment. *Soft Matter* **2010**, *6*, 4876–4884. [CrossRef]
38. Lv, P.F.; Wang, D.; Chen, Y.L.; Zhu, S.X.; Zhang, J.B.; Mao, L.K.; Gao, Y.X.; Yuan, F. Pickering emulsion gels stabilized by novel complex particles of high-pressure-induced WPI gel and chitosan: Fabrication, characterization and encapsulation. *Food Hydrocoll.* **2020**, *108*, 105992. [CrossRef]
39. Zhang, S.M.; Zhou, Y.H.; Yang, C. Pickering emulsions stabilized by the complex of polystyrene particles and chitosan. *Colloids Surf. A Physicochem. Eng. Asp.* **2015**, *482*, 338–344. [CrossRef]
40. Hou, Z.Q.; Zhang, M.; Liu, B.; Yan, Q.L.; Yuan, F.; Xu, D.X.; Gao, Y.X. Effect of chitosan molecular weight on the stability and rheological properties of beta-carotene emulsions stabilized by soybean soluble polysaccharides. *Food Hydrocoll.* **2012**, *26*, 205–211. [CrossRef]
41. Wu, T.T.; Huang, J.Q.; Jiang, Y.Y.; Hu, Y.Q.; Ye, X.Q.; Liu, D.H.; Chen, J.C. Formation of hydrogels based on chitosan/alginate for the delivery of lysozyme and their antibacterial activity. *Food Chem.* **2018**, *240*, 361–369. [CrossRef] [PubMed]
42. Niu, F.G.; Niu, D.B.; Zhang, H.J.; Chang, C.H.; Gu, L.P.; Su, Y.J.; Yang, Y.J. Ovalbumin/gum arabic-stabilized emulsion: Rheology, emulsion characteristics, and Raman spectroscopic study. *Food Hydrocoll.* **2016**, *52*, 607–614. [CrossRef]
43. Huang, Z.H.; Guo, B.Z.; Deng, C.; Tang, C.; Liu, C.M.; Hu, X.T. Fabrication and characterization of the W/O/W multiple emulsion through oleogelation of oil. *Food Chem.* **2021**, *358*, 129856. [CrossRef] [PubMed]

44. Cai, W.D.; Zhu, J.; Wu, L.X.; Qiao, Z.R.; Li, L.; Yan, J.K. Preparation, characterization, rheological and antioxidant properties of ferulic acid-grafted curdlan conjugates. *Food Chem.* **2019**, *300*, 125221. [CrossRef]
45. Mavrakis, C.; Kiosseoglou, V. The structural characteristics and mechanical properties of biopolymer/mastic gum microsized particles composites. *Food Hydrocoll.* **2008**, *22*, 854–861. [CrossRef]
46. Lu, X.X.; Zhang, H.W.; Li, Y.Q.; Huang, Q.R. Fabrication of milled cellulose particles-stabilized Pickering emulsions. *Food Hydrocoll.* **2018**, *77*, 427–435. [CrossRef]
47. Guo, J.; Zhou, Q.; Liu, Y.C.; Yang, X.Q.; Wang, J.M.; Yin, S.W.; Qi, J.R. Preparation of soy protein-based microgel particles using a hydrogel homogenizing strategy and their interfacial properties. *Food Hydrocoll.* **2016**, *58*, 324–334. [CrossRef]
48. Gasparov, V.S.; Degtyar, V.G. Protein Determination with Coomassie Brilliant Blue G-250. *Biochemistry* **1994**, *59*, 563–572.
49. Wang, L.J.; Yin, S.W.; Wu, L.Y.; Qi, J.R.; Guo, J.; Yang, X.Q. Fabrication and characterization of Pickering emulsions and oil gels stabilized by highly charged zein/chitosan complex particles (ZCCPs). *Food Chem.* **2016**, *213*, 462–469. [CrossRef]
50. Kholodenko, A.L. Statistical mechanics of the deformable droplets on Riemannian surfaces: Applications to reptation and related problems. *J. Math. Phys.* **1996**, *37*, 1314–1335. [CrossRef]
51. Chang, C.H.; Niu, F.G.; Gu, L.P.; Li, X.; Yang, H.; Zhou, B.; Wang, J.W.; Su, Y.J.; Yang, Y.J. Formation of fibrous or granular egg white protein microparticles and properties of the integrated emulsions. *Food Hydrocoll.* **2016**, *61*, 477–486. [CrossRef]
52. Wang, Z.G.; Zhang, N.; Chen, C.; He, R.; Ju, X.R. Rapeseed Protein Nanogels As Novel Pickering Stabilizers for Oil-in-Water Emulsions. *J. Agric. Food Chem.* **2020**, *68*, 3607–3614. [CrossRef] [PubMed]

molecules

MDPI

Article

Ultrasound-Assisted Preparation of Maillard Reaction Products Derived from Hydrolyzed Soybean Meal with Meaty Flavor in an Oil-In-Water System

Yongkang Ye [1,2], Shengquan Dai [1], Hongyan Zhang [1], Shudong He [1], Wanwan Hu [3], Xiaodong Cao [1,*] and Zhaojun Wei [1,2,*]

[1] School of Food Science and Biological Engineering, Hefei University of Technology, Hefei 230009, China
[2] Collaborative Innovation Center for Food Production and Safety, School of Biological Science and Engineering, North Minzu University, Yinchuan 750021, China
[3] Huangshan Chaogang Food Co., Ltd., Huangshan 245000, China
* Correspondence: xiaodongcao@hfut.edu.cn (X.C.); zjwei@hfut.edu.cn (Z.W.)

Abstract: In the present work, we prepared Maillard reaction products (MRPs) derived from enzyme hydrolyzed soybean meal with ultrasound assistance in an oil-(oxidized lard)-in-water system (UEL-MRPs) or oil-free system (UN-MRPs), and the effect of ultrasound on the properties of the obtained MRPs was evaluated. The analysis of fatty acids in lard with different treatments showed that ultrasound can generate more unsaturated fatty acids in the aqueous phase. The UV–Vis absorbances of UEL-MRPs, UN-MRPs, and MRPs obtained in an oil-in-water system (EL-MRPs) and MRPs obtained in an oil-free system (N-MRPs) at 294 and 420 nm indicated that ultrasound could increase the amount of Maillard reaction intermediates and melanoids in the final products of the Maillard reaction. This was in line with the result obtained from color change determination—that ultrasound can darken the resultant MRPs. Volatile analysis showed ultrasound can not only increase the number of volatile substances, but also greatly increase the composition of volatile substances in UEL-MRPs and UN-MRPs, especially the composition of those contributing to the flavor of the MRPs, such as oxygen-containing heterocycles, sulfur-containing compounds, and nitrogen-containing heterocycles. Descriptive sensory evaluation revealed that UN-MRPs and UEL-MRPs had the highest scores in total acceptance, ranking in the top two, and UEL-MRPs had the strongest meaty flavor among these four kinds of MRPs. Furthermore, the measurements of antioxidant activities, including DPPH radical-scavenging activity, hydroxyl radical scavenging ability, and ferric ion reducing antioxidant power, were conducted, showing that UN-MRPs exhibited the highest antioxidant activity among all the MRPs.

Keywords: Maillard reaction; ultrasound; hydrolyzed soybean meal; oxidized lard

Citation: Ye, Y.; Dai, S.; Zhang, H.; He, S.; Hu, W.; Cao, X.; Wei, Z. Ultrasound-Assisted Preparation of Maillard Reaction Products Derived from Hydrolyzed Soybean Meal with Meaty Flavor in an Oil-In-Water System. *Molecules* **2022**, *27*, 7236. https://doi.org/10.3390/molecules27217236

Academic Editor: Adele Papetti

Received: 5 September 2022
Accepted: 14 October 2022
Published: 25 October 2022

1. Introduction

Maillard reaction products (MRPs) have a great influence on food flavor. It can not only change the flavor and color of food [1], but also impart or enhance other properties of food, such as the antioxidant and antibacterial capacity of food [2]. There have been more and more studies on the preparation of meat flavors by using Maillard reaction [3,4]. Typically, the proteins involved in the Maillard reaction system are plant-derived proteins [5] and animal-derived proteins [1,6]. However, the meat flavor produced by only using plant-derived proteins in the Maillard reaction system has obvious deficiencies (such as light meat aroma) in meat flavor. Although the meat flavor prepared by the participation of animal-derived proteins in Maillard reaction has a strong meat flavor, the pretreatment process of animal-derived proteins is often complicated [7]. Studies have shown that oxidized animal fats have a strong meaty taste. Therefore, more and more research has been focused on improving the meat flavor of MRPs by adding different animal fats into the Maillard reaction system [3,8–11]. For example, Song et al. [12] treated lard with three different lipases and investigated the effect of lard treated with different lipases on MRPs in the xylose/glucose and cysteine Maillard reaction system. They concluded that the MRPs obtained by the addition of enzymatically hydrolyzed lard had better flavor, compared with the MRPs without lard addition. Moreover, the MRPs obtained by adding lard hydrolysate treated with lipase MER had the strongest meat flavor and the lowest off-flavor. The reasons for the positive effects of fat hydrolysates on the flavor of MRPs may include two aspects. One is that the substances, such as aldehydes, ketones, alcohols, acids, and lactones, produced by fat degradation have a certain positive impact on meaty taste [8]. The other is that the carbonyl compounds produced by fat oxidation can react with amino acids and reduce sugars and their related degradation products, producing some substances with lower odor thresholds, such as aliphatic alcohols and furans, which have a positive effect on the meaty taste enhancement of MRPs [13].

Ultrasound, which has been widely used in food industry, has been conducted on promotion of Maillard reaction in recent years [14–16]. This is because ultrasonic waves can not only accelerate the rate of the Maillard reaction and produce more intermediates and final products, but also improve the antioxidant properties of MRPs. There are three reasons for the promotion of the Maillard reaction by ultrasonic wave. First, the cavitation effect of ultrasonic waves can generate a transient high-temperature and high-pressure environment inside the liquid, which provides extreme reaction conditions for the Maillard reaction [17]. Second, the mechanical effect of the ultrasound can accelerate the mixing of the solution and increase the frequency of intermolecular collisions, leading to the acceleration of the reaction rate [15]. Third, the activation energy required for the ultrasound-induced Maillard reaction is lower than that of conventional heat treatment [16].

Soybean meal, the by-product of soybean oil extraction, which is usually used as animal feed, possesses a high protein content of 30–50%. In order to increase its added value, there are many studies on the application of soybean meal [18–20], including the preparation of Maillard flavor peptides from soybean meal enzymatic hydrolysis [4,5,21]. For instance, Yu et al. [5] used peptides of different molecular weights from soybean meal hydrolysis to participate in the Maillard reaction and investigated the relationship between the antioxidant and sensory properties of the MRPs and the molecular weight of the peptides. They found that the MRPs obtained from the Maillard reaction, participating via peptides with molecular weights of 1–3 kDa, showed strongest antioxidative properties, the highest umami, and the lowest bitterness taste. In present work, we prepared MRPs using soybean meal hydrolysates and oxidized lard in an oil-in water reaction system assisted with ultrasound. The effect of ultrasonic wave treatment on MRPs was evaluated by investigating the composition and types of flavor substances, the antioxidant properties, and the sensory properties of the obtained MRPs.

2. Results and Discussion

This section may be divided by subheadings. It should provide a concise and precise description of the experimental results, their interpretation, and the experimental conclusions that can be drawn.

2.1. Fatty Acids Composition Analysis

The types of fatty acids and their content in FL, EL, and UEL that determined by GC-MS are shown in Table 1. The fatty acids detected in FL, EL, and UEL mainly included nine saturated fatty acids and eight unsaturated fatty acids. As shown in the table, the highest content of saturated fatty acids is palmitic acid (C16:0), followed by stearic acid (C18:0), and the highest contents of unsaturated fatty acids are oleic acid (C18:1) and linoleic acid (C18:2). The total content of saturated fatty acids increased slightly from 46.016 ± 0.123% in FL to 46.814 ± 0.014% in EL, and then to 46.928 ± 0.136% in UEL. The total contents in saturated fatty acids in EL and UEL are slightly higher than that in FL, which indicates that enzymatic treatment and enzymatic treatment, followed with ultrasonication, can lead to the pyrolysis and oxidation of triglycerides to form more saturated fatty acids [4]. The displayed results in Table 1 also show that the total content of unsaturated fatty acid decreased slightly from 53.984 ± 0.123% in FL to 53.186 ± 0.014% in EL, and ultimately to 53.072 ± 0.136% in UEL. The tendency is consistent with the study by Yu et al. [22]. The reasons for the decrease of unsaturated fatty acid content may come from two aspects: First, ultrasound creates a relatively high temperature environment, which accelerates the oxidation of unsaturated fatty acids, thus reducing the content of unsaturated fatty acids in UEL. Second, under the action of ultrasonic waves, the substances in the system are fully mixed, which makes more unsaturated fatty acids migrate from the oil phase to the water phase, resulting in a decrease in the proportion of unsaturated fatty acids in the oil phase. The results indicate that more unsaturated fatty acids in the aqueous phase would participate in the Maillard reaction, which might lead to the difference in the meat aroma of UEL-MRPs and other different MRPs.

Table 1. GC results of fatty acid compositions in FL, EL, and UEL.

Fatty Acids	Percentage (%)		
	FL	EL	UEL
Saturated fatty acid			
C10:0 decanoic acid	0.020 ± 0.001 [a]	0.021 ± 0.000 [a]	0.017 ± 0.000 [b]
C12:0 lauric acid	0.076 ± 0.000 [a]	0.077 ± 0.003 [a]	0.075 ± 0.000 [a]
C14:0 myristic acid	0.898 ± 0.005 [a]	0.906 ± 0.005 [a]	0.904 ± 0.008 [a]
C15:0 pentadecanoic acid	0.041 ± 0.005 [a]	0.040 ± 0.002 [a]	0.041 ± 0.003 [a]
C16:0 palmitic acid	36.271 ± 0.090 [b]	36.861 ± 0.16 [a]	36.958 ± 0.142 [a]
C17:0 margaric acid	0.129 ± 0.012 [a]	0.130 ± 0.003 [a]	0.136 ± 0.006 [a]
C18:0 stearic acid	8.150 ± 0.020 [c]	8.345 ± 0.026 [b]	8.402 ± 0.005 [a]
C21:0 n-heneicosanoic acid	0.293 ± 0.007 [ab]	0.318 ± 0.026 [a]	0.279 ± 0.003 [b]
C22:0 behenic acid	0.137 ± 0.002 [a]	0.118 ± 0.000 [b]	0.116 ± 0.001 [b]
Total	46.016 ± 0.123 [b]	46.814 ± 0.014 [a]	46.928 ± 0.136 [a]
Unsaturated fatty acid			
C14:1 myristoleic acid	0.017 ± 0.001 [a]	0.017 ± 0.000 [a]	0.018 ± 0.005 [a]
C16:1 palmitoleic acid	1.066 ± 0.026 [a]	1.036 ± 0.001 [a]	1.046 ± 0.018 [a]
C17:1 heptadecenoic acid	0.060 ± 0.003 [a]	0.056 ± 0.005 [a]	0.058 ± 0.003 [a]
C18:1 oleic acid	37.328 ± 0.103 [a]	36.754 ± 0.018 [b]	36.722 ± 0.135 [b]
C18:2 linoleic acid	14.663 ± 0.027 [a]	14.441 ± 0.004 [b]	14.413 ± 0.009 [b]
C20:1 eicosenoic acid	0.185 ± 0.005 [b]	0.211 ± 0.003 [a]	0.164 ± 0.016 [c]
C18:3 α-linolenic acid	0.633 ± 0.003 [a]	0.638 ± 0.015 [a]	0.626 ± 0.001 [a]
C20:3 carbonium	0.033 ± 0.003 [a]	0.033 ± 0.000 [a]	0.025 ± 0.001 [b]
Total	53.984 ± 0.123 [a]	53.186 ± 0.014 [b]	53.072 ± 0.136 [b]

Note: Results were expressed as mean value ± standard deviation ($n = 3$). Values bearing different letters (a to c) were significantly different ($p < 0.05$).

2.2. Browning Intensity of the MRPs

The degree of browning of the MRPs usually changes as the Maillard reaction proceeds. The UV–Vis absorbance at 294 nm is typically adopted to monitor the formation of Maillard reaction intermediates, while the absorbance at 420 nm is used to evaluate the brown polymer in the final products [23]. As shown in Figure 1, compared with the MRPs (N-MRPs and EL-MRPs) prepared without ultrasound assistance, the ultrasound-assisted MRPs (UN-MRPs and UEL-MRPs) exhibited higher absorbance at 294 nm (Figure 1A), and the UV absorption at 420 nm (Figure 1B) also showed the same trend as that at 294 nm. This indicates that ultrasound can accelerate the formation of Maillard reaction intermediates and increase the brown polymer in the final products. The reasons for this result may be related to two aspects. On the one hand, the mechanical effect of ultrasound can accelerate the mixing of solutions and increase the frequency of collisions between molecules [24]. This enables more Amadori compounds to degrade into Maillard reaction intermediates, which increases the absorbance of MRPs at 294 nm; at the same time, these intermediate products can be further converted into melanoid substances, thereby increasing the absorbance of

MRPs at 420 nm. On the other hand, due to the cavitation effect of ultrasound, a transient high-temperature and high-pressure environment can be generated inside the liquid [17]. Compared with traditional heating procedure, ultrasound can positively promote the Maillard reaction, resulting in the production of more intermediates and melanoids, leading to higher A294/420 of the ultrasound-assisted MRPs. In addition, compared with the MRPs (N-MRPs and UN-MRPs) without lard participating in the Maillard reaction, the MRPs (EL-MRP and UEL-MRP) obtained by EL participating in the Maillard reaction showed higher absorbance at 294 and 420 nm. This may be due to the aldehydes and ketones produced by the oxidation of lard participating in the Maillard reaction, resulting in more Maillard reaction intermediates and final products [8].

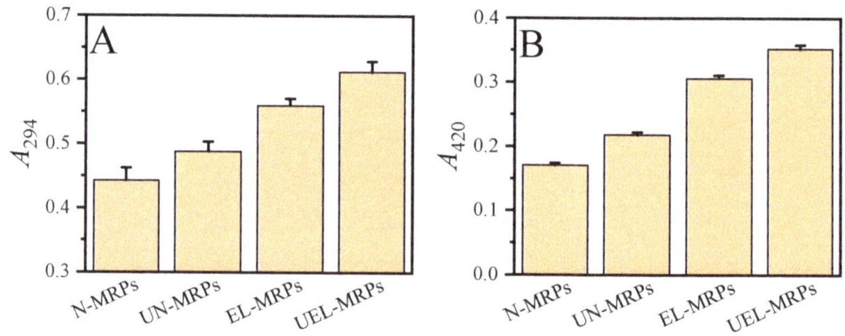

Figure 1. The UV–Vis absorbance at 294 nm (**A**) and 420 nm (**B**).

2.3. Changes in Color

The measured values of ΔL^*, Δa^*, Δb^* reflecting the color changes of the MRPs are shown in Table 2.

Table 2. Color changes of various MRPs.

Sample	ΔL^*	Δa^*	Δb^*	ΔE^*
N-MRPs	-2.61 ± 0.06 [a]	-0.06 ± 0.02 [a]	-2.07 ± 0.06 [c]	3.33 ± 0.02 [c]
UN-MRPs	-2.68 ± 0.03 [a]	-0.14 ± 0.02 [c]	-2.21 ± 0.01 [d]	3.47 ± 0.02 [b]
EL-MRPs	-3.42 ± 0.08 [b]	-0.11 ± 0.02 [b]	-1.60 ± 0.03 [b]	3.78 ± 0.09 [a]
UEL-MRPs	-3.53 ± 0.02 [c]	-0.08 ± 0.01 [ab]	-1.53 ± 0.02 [a]	3.85 ± 0.02 [a]

Note: Results were expressed as mean value ± standard deviation ($n = 3$). Values bearing different letters (a to c) were significantly different ($p < 0.05$).

It is clear that all these parameters are negative, compared to distilled water, indicating that the color of the MRPs appears darker, greener, and bluer. In addition, compared with N-MRPs and EL-MRPs, the ΔL^* values of UN-MRPs and UEL-MRPs were relatively lower. This demonstrates that the color of the MRPs obtained with ultrasound assistance were darker than those obtained without the assistance of ultrasound. This further suggests that ultrasound treatment can accelerate the Maillard reaction process, resulting in more melanoids in the MRPs [1,25]. It can also be seen from the table that the ΔL^* values of EL-MRPs and UEL-MRPs were higher than those of N-MRPs and UN-MRPs. This may be due to the fact that fat degradation products participate in the Maillard reaction, producing more dicarbonyl compounds and melanoids, thus changing the color of the MRPs [3].

The ΔE^* values of these MRPs are also displayed in Table 2. It is easy to find that the ΔE^* value of UEL-MRPs is the largest among these MRPs, which may be related to the content of low molecular weight polymers [4,26].

2.4. GC–MS/SPME Analysis of the Volatile Components in the MRPs

The volatile components in the MRPs were identified by GC-MS/SPME, and the identified substances and their corresponding contents are listed in Table 3. As listed in the table, a total of 49 volatile compounds were identified in these MRPs, including 7 aldehydes, 4 ketones, 11 alcohols, 3 esters, 9 acids, 3 hydrocarbons, 4 phenols, 1 ether, 2 pyrazines, 1 pyrrole, 2 furans, 1 thiazole, and 1 thiophene. Due to the differences in the presence/absence of oxidized lard participating in the reaction and with/without the ultrasound-assisted reaction in the implementation of Maillard reaction, the types and contents of volatile components in the different MRPs were not the same. The number of the detected substances in UEL-MRPs, UN-MRPs, EL-MRPs, and N-MRPs were 35, 32, 29, and 26, respectively. Among these volatile compounds, the identified compounds, such as oxygen-containing heterocycles (furans), sulfur-containing heterocycles (thiophenes and thiazoles), and nitrogen-containing heterocycles (pyrazines and pyrroles) have a greater impact on the flavor of MRPs [4].

Table 3. Volatile compounds in the MRPs.

No.	Volatile Compounds	[1] KIs	[2] Odors	Relative Concentration [ng kg^{-1}] (Mean ± SD)			
				N-MRPs	UN-MRPs	EL-MRPs	UEL-MRPs
	Aldehydes (7)			16,325.90 ± 2633.65 ^{b}	23,312.20 ± 2256.92 ^{b}	42,201.86 ± 6487.71 ^{a}	51,094.08 ± 6451.28 ^{a}
1	2-undecenal	1311	waxy	—	—	—	1526.60 ± 439.78
2	nonanal	1104	fatty, citrus	2544.39 ± 789.48 ^{c}	3150.21 ± 1627.73 ^{c}	13,575.11 ± 744.44 ^{a}	8164.12 ± 1311.46 ^{b}
3	octanal	92	fatty, citrus, honey	—	628.01 ± 58.54 ^{c}	5827.26 ± 424.38 ^{a}	2740.49 ± 624.90 ^{b}
4	benzaldehyde	982	almond	13,383.74 ± 2478.88 ^{b}	18,797.82 ± 385.02 ^{b}	21,529.94 ± 5270.65 ^{a}	38,662.87 ± 8787.69 ^{a}
5	4-methoxy-benzaldehyde	1171	hawthorn	—	214.49 ± 59.06	—	—
6	decanal	1204	fatty, sweet orange	397.76 ± 65.39 ^{b}	521.67 ± 154.22 ^{b}	868.86 ± 65.75 ^{a}	—
7	(E)-2-octenal	1013	—	—	—	400.68 ± 178.09	—
	Ketones (4)			500.64 ± 33.39 ^{b}	592.36 ± 40.27 ^{b}	635.48 ± 117.50 ^{b}	4302.12 ± 1105.77 ^{a}
8	2H-pyran-2,6(3H)-dione	1098	—	91.11 ± 20.19 ^{b}	135.62 ± 13.81 ^{b}	140.31 ± 40.39 ^{b}	386.13 ± 76.23 ^{a}
9	acetoin	717	buttery	112.13 ± 23.42 ^{c}	106.88 ± 20.41 ^{c}	495.16 ± 77.50 ^{b}	756.99 ± 160.41 ^{a}
10	1-hydroxy-2-propanone	698	—	297.40 ± 9.22	—	—	3159.00 ± 875.38
11	6-methyl-5-hepten-2-one	938	fatty, green, citrus-like	—	349.86 ± 48.08	—	—
	Alcohols (11)			4696.10 ± 431.42 ^{d}	10,404.52 ± 1020.19 ^{c}	13,917.84 ± 1352.76 ^{b}	37,637.68 ± 2448.57 ^{a}
12	2-furanmethanol	885	burnt, caramel	951.36 ± 27.19 ^{b}	1242.79 ± 330.92 ^{b}	2623.38 ± 348.82 ^{a}	1000.58 ± 152.03 ^{b}
13	1-pentanol	761	—	—	621.58 ± 175.56 ^{c}	4269.34 ± 411.52 ^{b}	9280.11 ± 316.91 ^{a}
14	1-octen-3-ol	969	mushroom	—	—	—	7635.86 ± 1535.68
15	2-methyl-3-pentanethiol	793	—	—	2140.45 ± 320.99	—	1730.77 ± 342.90
16	1-hexanol	860	green, fruity	1106.16 ± 39.61 ^{c}	2882.99 ± 293.59 ^{b}	1959.74 ± 285.49 ^{bc}	6247.56 ± 933.66 ^{a}
17	1,4-butanediol	904	—	112.70 ± 27.90	—	—	424.19 ± 15.77
18	benzyl alcohol	1036	fruity	—	—	1134.57 ± 175.23	—
19	phenylethyl alcohol	1136	roses	529.12 ± 17.16	649.71 ± 92.40	—	—
20	1-heptanol	960	weak alcoholic	—	—	—	4115.44 ± 414.03
21	maltol	1063	caramel	1559.63 ± 345.82 ^{c}	2331.50 ± 259.26 ^{c}	3930.81 ± 601.33 ^{b}	7203.18 ± 538.75 ^{a}
22	1-dodecanol	1457	fatty	437.12 ± 145.13	535.49 ± 89.77	—	—
	Esters (3)			1413.30 ± 226.14 ^{c}	1755.50 ± 99.63 ^{bc}	2245.81 ± 515.19 m^{b}	5204.68 ± 231.95 ^{a}
23	butyrolactone	825	—	1003.15 ± 223.50 ^{c}	1192.48 ± 212.87 ^{c}	2245.81 ± 630.98 ^{b}	3749.90 ± 193.20 ^{a}
24	5-ethyldihydro-2(3H)-furanone	986	caramel	—	—	—	573.71 ± 44.52
25	hexadecanoic acid, methyl ester	1878	—	410.15 ± 12.07 ^{b}	563.03 ± 154.90 ^{b}	—	881.07 ± 9.14 ^{a}
	Acids (9)			8519.72 ± 535.84 ^{b}	10,217.13 ± 741.44 ^{b}	15,040.33 ± 997.20 ^{a}	14,338.67 ± 1294.21 ^{a}
26	isovaleric acid	811	rancid	5806.91 ± 786.41 ^{b}	5495.16 ± 592.05 ^{b}	8090.93 ± 853.77 ^{a}	7441.90 ± 54.91 ^{a}
27	n-decanoic acid	1372	fatty, rancid	—	378.11 ± 26.06	721.44 ± 142.74	—
28	hexanoic acid	974	fatty, waxy,	1480.04 ± 399.09 ^{c}	1781.70 ± 230.72 ^{bc}	2802.08 ± 545.34 ^{a}	2394.52 ± 288.82 ^{ab}
29	octanoic acid	1173	waxy, fatty	476.11 ± 97.97 ^{c}	1156.04 ± 155.08 ^{b}	2114.29 ± 498.28 ^{a}	2560.38 ± 451.35 ^{a}
30	nonanoic acid	1272	—	264.56 ± 8.58 ^{c}	378.17 ± 62.92 ^{b}	461.28 ± 40.42 ^{a}	485.55 ± 17.18 ^{a}
31	heptanoic acid	1073	waxy, fruity, fatty	226.64 ± 24.61 ^{c}	421.16 ± 68.00 ^{b}	538.76 ± 96.81 ^{b}	673.93 ± 65.34 ^{a}
32	pentanoic acid	875	—	139.37 ± 23.12	606.79 ± 71.25	—	—
33	butanoic acid	811	rancid	—	—	311.54 ± 20.75	—
34	pentadecanoic acid	1869	waxy	126.09 ± 34.85	—	—	—

Table 3. *Cont.*

No.	Volatile Compounds	[1] KIs	[2] Odors	Relative Concentration [ng kg^{-1}] (Mean ± SD)			
				N-MRPs	UN-MRPs	EL-MRPs	UEL-MRPs
	Pyrazines (2)			—	756.89 ± 144.74	—	2235.33 ± 311.44
35	2,5-dimethyl-pyrazine	894	Cocoa, roasted, nutty	—	—	—	2235.33 ± 311.44
36	tetramethyl-pyrazine		nutty, chocolate, coffee	—	756.89 ± 144.74	—	—
	Furans (2)			600.96 ± 73.22 ^{b}	842.49 ± 191.62 ^{c}	2369.50 ± 398.75 ^{a}	2565.21 ± 176.92 ^{a}
37	bis(2-methyl-3-furyl) disulfide	1745	roasty, meat, sulfur	600.96 ± 73.22 ^{b}	842.49 ± 191.62 ^{a}	666.67 ± 91.68 ^{ab}	524.40 ± 94.07 ^{b}
38	1-(2-furanyl)-Ethanone	878	almond, nut, roasted	—	—	1702.83 ± 329.22	2040.81 ± 89.87
	Hydrocarbons (3)			—	678.87 ± 208.19 ^{ab}	1047.89 ± 327.45 ^{a}	314.59 ± 50.54 ^{b}
39	n-hexane	618	gasoline	—	—	—	314.59 ± 50.54
40	nonadecane	1910	—	—	678.87 ± 208.19	—	—
41	pentadecane	1512	—	—	—	1047.89 ± 327.45	—
	Phenols (4)			1094.01 ± 59.50 ^{d}	2181.28 ± 357.30 ^{c}	6203.81 ± 445.34 ^{b}	7909.26 ± 221.45 ^{a}
42	butylated hydroxytoluene	1668	—	—	—	3796.72 ± 348.50	3963.23 ± 502.73
43	phenol	901	—	395.39 ± 30.41 ^{c}	378.72 ± 92.58 ^{c}	636.09 ± 131.79 ^{b}	1090.62 ± 146.61 ^{a}
44	p-cresol	1014	—	—	526.95 ± 189.71	—	695.62 ± 166.81
45	2-methoxy-phenol	1090	—	698.62 ± 65.46 ^{d}	1275.61 ± 247.70 ^{c}	1770.99 ± 41.55 ^{b}	2159.78 ± 92.60 ^{a}
	Thiazoles (1)						
46	4-methyl-5-thiazoleethanol	1264	meaty, roasted	1582.85 ± 236.87 ^{c}	2018.86 ± 74.48 ^{c}	4492.16 ± 215.17 ^{b}	8787.13 ± 1646.45 ^{a}
	Ethers (1)						
47	3-tert-butyl-4-hydroxyanisole	1417	—	—	—	752.29 ± 76.05	742.80 ± 18.10
	Pyrroles (1)						
48	1-(1H-pyrrol-2-yl)-ethanone	1035	walnuts, toast	237.32 ± 5.26 ^{d}	477.30 ± 83.31 ^{c}	665.75 ± 10.75 ^{b}	1288.37 ± 146.92 ^{a}
	Thiophenes (1)						
49	5-methyl-2-thiophenecarboxaldehyde	1072	almond, fruity, nutty	—	—	—	873.56 ± 95.92

Note: Means bearing different letters are significantly ($p < 0.05$) different in the same line. "—", not detected. [1] KI (Kovats indices) determined by searching the mass spectrum in the database and manual interpretation. [2] Odors indicated the odor bias of some specific flavor compound.

The histogram of the content of oxygen-containing compounds in these MRPs is shown in Figure 2A. As shown in the figure, oxygen-containing compounds are the most abundant in the types and contents of volatile compounds in all MRPs. Moreover, the oxygen-containing volatile compounds contents in EL-MRP and UEL-MRP were significantly higher than that in N-MRPs and UN-MRPs. This suggests that the participation of oxidized lard in the hydrolyzed soybean-based Maillard reaction system can significantly increase the content of oxygen-containing compounds in MRPs volatile compounds, which is similar to the results of the reported research [3]. In addition, the contents of oxygen-containing volatile compounds in UN-MRPs and UEL-MRPs were higher than those in N-MRPs and EL-MRPs, respectively. This may be related to the degree of Maillard reaction.

Furans, a class of oxygen-containing volatile compounds, have a greater impact on the flavor of MRPs [27]. Figure 2B displays the amounts of the identified furans (light cyan histogram) and a specific compound 1-(2-furanyl)-ethanone (magenta histogram) in the MRPs, respectively. It is obvious that the total amount of furans increased in the order of N-MRPs, UN-MRPs, EL-MRPs, and UEL-MRPs, and only 1-(2-furanyl)-ethanone was detected in EL-MRPs and UEL-MRPs. The 1-(2-furanyl)-ethanone has sweet, cocoa, almond, and caramel flavors [28], which can improve the flavor of MRPs. Furans can be generated from fatty acid oxidation or from glycerol and cysteine degradation products through a complex series of chemical reactions (cyclization, dehydration, and aldol condensation) [10]. Due to the participation of lard in the Maillard reaction, coupled with the assistance of ultrasound, the furan content in UN-MRPs, EL-MRPs, and UEL-MRPs was higher than that in N-MRPs, which might exert a positive effect on the flavor of these MRPs.

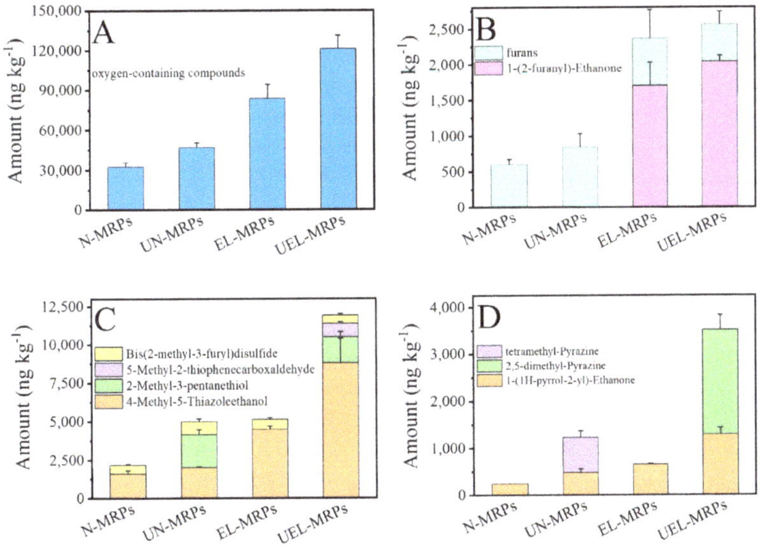

Figure 2. (**A**) The total amount of oxygen-containing compounds in various MRPs. (**B**) The amount of furans and 1-(2-furanyl)-ethanone in various MRPs. (**C**) The total amount of sulfur-containing compounds in various MRPs. (**D**) The total amount of nitrogen-containing compounds in various MRPs.

Sulfur-containing compounds generally have a lower odor threshold and show great impact on the meaty aroma of different foods [29]. As shown in Figure 2C, a total of four sulfur-containing substances were detected in these MRPs, namely 2-methyl-3-pentanethiol, 4-methyl-5-thiazoleethanol, bis(2-methyl-3-furyl)disulfide, and 5-methyl-2-thiophenecarboxaldehyde. It is obvious that the total amount of sulfur containing compounds in UEL-MRPs (11,915.86 ng kg^{-1}) was higher than that in EL-MRPs (5158.83 ng kg^{-1}), and in UN-MRPs (5001.8 ng kg^{-1}), it was higher than that in N-MRP (2183.81 ng kg^{-1}). Moreover, the MRPs with oxidized lard participation in the Maillard reaction had higher content of sulfur containing compounds than the MRPs obtained without the presence of oxidized lard. Another result that should be mentioned is that 5-methyl-2-thiophenecarboxaldehyde, which contributes to meat flavor [30], was found only in UEL-MRPs (873.56 ng kg^{-1}). The identification and quantitation results of the sulfur-containing substances might be due to the participation of animal fat in the Maillard reaction [4] and the promotion of ultrasound to the Maillard reaction [15].

Nitrogen-containing heterocycles are another class of volatile substances contributing to the MRPs flavor. The nitrogen-containing heterocycles mostly exist in the forms of pyrazine, pyrrole, and pyridine. As shown in Figure 2D, altogether, three kinds of such substances were identified in these MRPs, namely tetramethyl-Pyrazine, 2,5-dimethyl-pyrazine, and 1-(1H-pyrrol-2-yl)-ethanone. The total contents of nitrogen-containing heterocycles in UN-MRPs (1234.19 ng kg^{-1}) and UEL-MRPs (3523.7 ng kg^{-1}) were higher than those in N-MRPs (237.32 ng kg^{-1}) and EL-MRPs (665.75 ng kg^{-1}), respectively. This may be due to the accelerated formation of nitrogen-containing compounds by using ultrasound assistance. Pyrazine generally has a nutty and cooked burnt aroma [31]. Figure 2D also shows that 2,5-dimethylpyrazine (2235.33 ng kg^{-1}) and tetramethyl-Pyrazine (756.89 ng kg^{-1}) were detected only in UEL-MRPs and in UN-MRP, respectively. This may be due to the fact that the activation energy values required for the synthesis of targeted pyrazine species in an ultrasound-assisted Maillard reaction model system were lower than those in heat treatment [32].

2.5. Sensory Evaluation

The sensory scoring criteria are listed in Table 4, and the sensory evaluation results of the MRPs are shown in Figure 3. It is clear that the participation of oxidized lard in the Maillard reaction and the application of ultrasound made little difference, regarding the salty and umami taste of these MRPs. The meaty taste of EL-MRPs and UEL-MRPs obtained from oxidized lard participation in the Maillard reaction showed better meat flavor than that of the N-MRPs and UN-MRPs obtained from the absence of oxidized lard in the Maillard reaction. In addition, the meat taste scores of the UEL-MRPs and UN-MRPs obtained by the promotion of ultrasound to the Maillard reaction were also higher than those of the EL-MRPs and N-MRPs, respectively. These may be related to their higher content of sulfur-containing compounds and nitrogen-containing heterocycles, such as thiophene and thiazole [13]. However, oxidized lard developed some undesired flavors during oxidation, which made EL-MRP and UEL-MRP have a heavier off-flavor. Compared with the MRPs of N-MRP and EL-MRP, UN-MRP and UEL-MRP had higher scores, in terms of total acceptance. The sensory evaluation results indicate that the flavor of MRPs could be improved by the application of ultrasound.

Table 4. Sensory evaluation scoring criteria of MRPs.

Sensory Indicators	Judging Controls	Scoring Criteria/Point
Off-flavor	The unaccepted flavor of rotten eggs, prepared by putting broken eggs (100 g) at 50 °C for 7 days, was used as odor intensity evaluation.	Strong odor: 0–2 Medium odor: 2–5 Lighter odor: 5–7 Odorless: 7–10
Meaty	Take certain pork lean meat, cut into 2.5 cm cubes, cook in water for 2 h, and then use as a meat flavor evaluation control.	Strong odor: 7–10
Umami	The umami used sodium glutamate solution (1%, w/v) as the umami note.	Medium odor: 5–7
Salty	Salty taste is the taste of 0.5% (w/v) sodium chloride solution.	Lighter odor: 2–5
Total acceptance	Evaluation based on meaty, umami, salty, and off-flavor.	Odorless: 0–2

Note: The score was given on a scale of 0 (undetected) to 10 (strong). The sensory evaluation standard of off-flavor is opposite to other indexes.

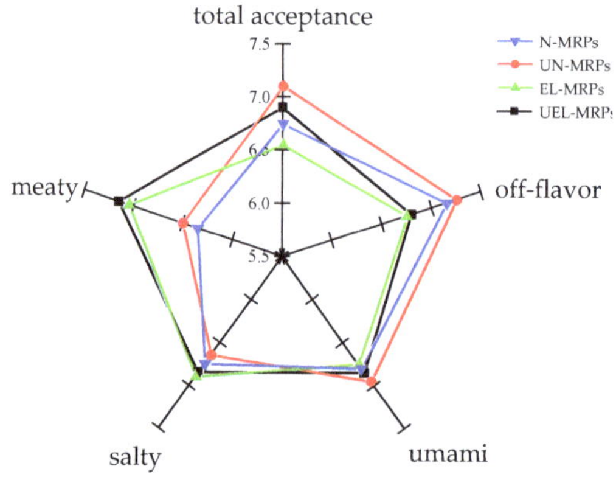

Figure 3. Sensory characteristics of various MRPs.

2.6. Sensory Evaluation

The antioxidant ability of the MRPs was evaluated by measuring the DRS, HSR, and ferric ion reducing antioxidant power. Figure 4A shows the DPPH radical scavenging ability of the MRPs. Under the same condition, the scavenging rate of DPPH radicals by UN-MRPs was the strongest among the four MRPs, and the scavenging rate reached a peak value of about 94.58%, at a concentration of 1 mg mL^{-1}. Furthermore, basically the scavenging rate of the DPPH radicals by these MRPs was in the order of UN-MRPs > N-MRPs > EL-MRPs > UEL-MRPs. As can be seen in Figure 4B,C, the hydroxyl radical scavenging ability (Figure 4B) and ferric ion reducing antioxidant power (Figure 4C) of the MRPs showed similar trends to the scavenging rates for DPPH radicals.

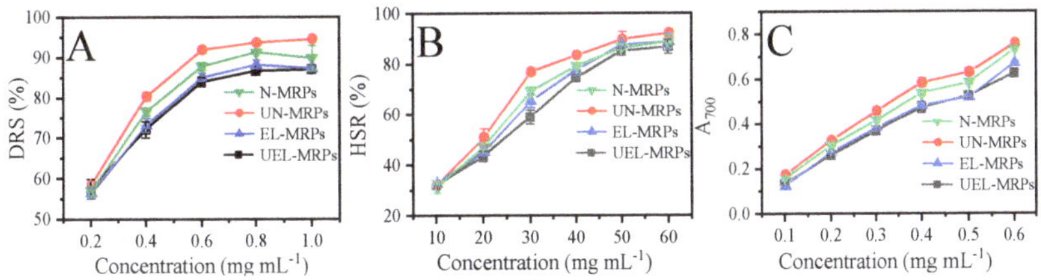

Figure 4. Antioxidant activity of various MRPs. (**A**) DPPH free radical scavenging activity. (**B**) Hydroxyl radical scavenging ability. (**C**) Reducing power, expressed in absorbance read at 700 nm.

According to the antioxidant analysis, we can find that UN-MRPs showed the best antioxidant ability among these MRPs. This may be because the ultrasound-assisted heat treatment accelerates the Maillard reaction, thereby generating more substances with antioxidant capacity [1]. It has been reported that the produced intermediate pyrroles and melanoids in the final products of the Maillard reaction have the antioxidant capacity [15,33]. According to the results of the browning intensity measurement and SPME/GC-MS, more intermediate substances, melanoids, and nitrogen-containing heterocycles were found in the UEL-MRPs. However, the antioxidant capacity of UEL-MRP was measured to be the lowest among these MRPs, and that of EL-MRPs was the second lowest. Studies have shown that fragments of 30–50 kDa in the MRPs have high antioxidant activity [34,35]. The possible reason for the difference in the antioxidant activity of the MRPs may be due to the distribution of the molecular weights in these MRPs. Future work will focus on obtaining MRPs with different molecular weights and evaluating their antioxidant activity, thereby finding out the relationship between the molecular weight and antioxidant activity.

3. Materials and Methods

3.1. Materials and Chemicals

Soybean meal was purchased from Muge Feed Co., Ltd. (Hebei, China). Lard was from Yusheng Edible Oil Co., Ltd. (Shandong, China). Lipase MER (7500 Lu g^{-1}) was purchased from Tianye Enzyme Preparation Co., Ltd. (Jiangsu, China). Alkaline protease (200,000 U g^{-1}), neutral protease (100,000 U g^{-1}), and flavor protease (200,000 U g^{-1}) were obtained from Shanghai Yuanye Biotechnology Co., Ltd. (Shanghai, China). L-cysteine, D-xylose, and thiamine were purchased from Shanghai Aladdin Biochemical Technology Co., Ltd. (Shanghai, China). Other chemicals were analytical reagents and purchased from Shanghai Mackin Biochemical Co., Ltd. (Shanghai, China). Deionized water was used throughout the experiment.

3.2. Sample Preparation

3.2.1. Preparation of Enzymatic Hydrolyzed Lard

Lipase MER was selected to prepared hydrolyzed lard [12]. Fresh lard (FL) and phosphate buffer (pH = 6.0) were mixed thoroughly at a ratio of 1:1 (*w*/*w*) and then added into a round-bottomed flask, followed with the addition of lipase MER (1 g enzyme per 100 g lard). The enzymatic hydrolysis was carried out at 45 °C for 1.5 h with magnetic stirring at a speed of 150 rmp. The enzymatic hydrolysis solution was then heated to 95 °C for 15 min to inactivate the lipase MER. The resultant solution was cooled down to room temperature, followed with centrifugation at 4000 rpm for 15 min. The supernatant solution, denoted as EL, was collected and kept at −20 °C for later use. The EL solution, treated with ultrasonication, was prepared by putting the probe of the ultrasound device (JY98-IIIDN, Ningbo Xinzhi Biotechnology Co., Ltd., Ningbo, China) 1.5 cm below the liquid level of the above collected supernatant solution (mixed solution of soybean meal enzymatic hydrolyzate, xylose, cysteine, and EF). The solution was sonicated at 300 W (5 s-on-5 s-off) for 30 min, and the obtained treated lard was named UEL. The acid values (AV) of FL, EL, and UEL were measured to be 0.48 ± 0.04, 61.84 ± 3.62, and 69.60 ± 1.73 mg g^{-1}, respectively; the peroxide values (PV) were 0.080 ± 0.004, 0.085 ± 0.003, and 0.084 ± 0.005 meq. kg^{-1}, respectively; and the p-anisidine values (p-AV) were 3.77 ± 0.43, 5.13 ± 0.53, and 6.04 ± 0.69, respectively.

3.2.2. Preparation of Soybean Meal Hydrolysates

Soybean meal hydrolysates were prepared according to the method of Zhang et al. [21], with tiny modifications. Soybean meal (47.8% of protein content) was pulverized and sieved with a 100-mesh sieve. The soybean meal powder was then dispersed in water to prepare suspension with a protein content of 4% (*w*/*w*). It was then heated to 90 °C for 30 min, cooled to room temperature, and then sonicated at 45 °C for 20 min with an ultrasonic power of 350 W. The suspension was then enzymatically hydrolyzed by a two-step enzymatic hydrolysis method. Firstly, the pH of the suspension was adjusted to 10.0, with 6.0 mol L^{-1} of NaOH, and then it was enzymatically hydrolyzed with alkaline protease (8000 U g^{-1}) at 50 °C for 3.5 h. Secondly, the pH of the above mixture was adjusted to 6.5, with 6 mol L^{-1} HCl, and it was enzymatically hydrolyzed with neutral protease (8000 U g^{-1}) and flavor protease (1200 U g^{-1}) at 45 °C for 4 h. The enzymatic hydrolysis solution obtained in each enzymatic hydrolysis step should be inactivated at 90 °C for 15 min, and then the soybean meal enzyme hydrolysate was centrifuged at 8000 rmp for 20 min. Finally, the supernatant was collected and stored at −20 °C for later use. The degree of hydrolysis of soybean meal enzymolysis solution was 39.04%, measured by using ortho-phthalaldehyde method.

3.2.3. Preparation of Maillard Reaction Products

The Maillard reaction system without the addition of lard was prepared by adding cysteine (2.0%), xylose (3.0%), and VB_1 (0.05%) into the soybean meal enzymatic hydrolysis, and the system with the addition of EL was prepared by adding cysteine (2.0%), xylose (3.0%), VB_1 (0.05%), and EL (1.0%) into the soybean meal enzymatic hydrolysis. The prepared reaction solution was mixed thoroughly by stirring at 50 °C for 10 min, with a stirring speed of 150 rpm. The preparation of MRPs with the assistance of ultrasound was carried out in accordance with the following steps. First, the pH of the reaction solution was adjusted to 7.1. Next, it was sonicated by putting the probe of the ultrasound device 1.5 cm below the liquid level. The solution was sonicated at 300 W (5 s-on-5 s-off) for 30 min. The Maillard reaction was performed by maintaining the above solution at 120 °C for 2 h, with constantly stirring at a speed of 200 rpm. Finally, the obtained MRPs were rapidly cooled in an ice-water bath, and then freeze-dried to prepare a lyophilized powder, which was stored at −20 °C for later use. The obtained MRPs, with and without the addition of EL, were named as UEL-MRPs and UN-MRP, respectively.

The procedure for the preparation of MRPs without ultrasound assistance was similar to the MRPs prepared with the assistance of ultrasound, except that no sonication step was required. The obtained MRPs with and without the addition of EL were named EL-MRPs and N-MRP, respectively.

3.3. *Analysis Methods*

3.3.1. Analysis of the Fatty Acid Composition in Various Lard

The fatty acid compositions in FL, EL, and UEL were analyzed using gas chromatography (GC-2010 PRO, Excellence in Science, Inc, Tokyo, Japan), equipped with a chromatographic column of SP-2560 (100 m × 0.25 mm × 0.20 μm). Before analyzing, FL, EL, and UEL were methylated using methanol containing 1 mol L^{-1} NaOH. Nitrogen was used as the carrier gas, and the gas flow was 1.0 mL min^{-1}. The temperature program settings were in accordance with the method of Ye et al. [4].

3.3.2. Determination of Browning Intensity and Color of MRPs

The browning intensity of these MRPs was measured on a UV-4802 UV–Vis spectrophotometer (Shanghai Unico Instrument Co., Ltd., Shanghai, China). The MRPs were diluted 150-fold and 60-fold with distilled water, respectively, and the absorbance values were measured at 294 nm (150-fold) and 420 nm (60-fold) using a UV-4802 UV–Vis spectrophotometer.

The color of these MRPs was determined by measuring the Commission International Eclairage (CIE) of lightness (L), parameter a (redness or greenness), and parameter b (yellowness or blueness). The measurements were performed on a NR200 portable colorimeter (Shenzhen, China) by recording ΔL^*, Δa^*, and Δb^*, and the total color difference (ΔE) was calculated according to the following equation [3]:

$$\Delta E = \sqrt{(\Delta L^*)^2 + (\Delta a^*)^2 + (\Delta b^*)^2} \quad (1)$$

3.3.3. Determination of DPPH Radical-Scavenging Activity

The DPPH radical-scavenging ability of MRPs was determined according to the method of Yu et al. [5] and Zeng et al. [36], with slight modifications. Typically, 100 μL of the MRPs solution (0.2, 0.4, 0.6, 0.8, and 1.0 mg mL^{-1}) and 100 μL DPPH (0.2 mmol L^{-1} in ethanol) solution were added dropwise to a 96-well microtiter plate. All these mixtures were then placed in the dark at 25 °C for 10 min. The absorbance of each sample was measured at 517 nm on a microplate reader (BioTek Instruments, Inc, Winooski, VT, USA), and the value was defined as A_S. Meanwhile, the absorbance of ethanol (200 μL) and the mixture of ethanol (100 μL) and the MRPs (100 μL) were also recorded, and the values were defined as A_c (absorbance of control) and A_b (absorbance of blank). The DPPH radical-scavenging activity (DRS%) of the MRPs was then calculated according to Equation (2), as follows:

$$\text{DRS\%} = \left(1 - \frac{A_S - A_c}{A_b}\right) \times 100 \quad (2)$$

3.3.4. Determination of Hydroxyl Radical Scavenging Ability

The determination of hydroxyl radical scavenging ability of the MRPs referred to the method described by Li et al. [37]. Briefly, 100 μL of the MRPs with different concentrations (10, 20, 30, 40, 50, and 60 mg mL^{-1}), 100 μL of salicylic acid-ethanol solution (10 mmol L^{-1}), 100 μL of $FeSO_4$ (10 mmol L^{-1}) solution, 700 μL of distilled water, and 1.0 mL of hydrogen peroxide solution (100 mmol L^{-1}) were thoroughly mixed and incubated at 37 °C for 15 min. Then, 250 μL of the above solution was added into a 96-well microtiter plate, and the absorbance of the sample was measured at 510 nm with a microplate reader. The control and blank samples were prepared by replacing salicylic acid-ethanol and the MRPs sample with ultrapure water, respectively. The hydroxyl radical scavenging capacity (HRS%) was calculated, according to Equation (3), as:

$$\text{HRS\%} = \left(1 - \frac{A_S - A_c}{A_b}\right) \times 100 \quad (3)$$

where A_S, A_c, and A_b are the absorbance of the sample, control, and blank sample, respectively.

3.3.5. Determination of Ferric Ion Reducing Ability

The reducing ability of MRPs was evaluated with reference to the method of Habinshuti et al. [2]. A volume of 2.0 mL of the MRPs with various concentrations (0.1, 0.2, 0.3, 0.4, 0.5, and 0.6 mg mL^{-1}), 2.5 mL of phosphate buffer saline (PBS, 0.2 mol L^{-1}, pH = 6.6), and 2.5 mL of $K_3Fe(CN)_6$ solution (1.0%, *w*/*v*) were mixed and incubated at 50 °C for 20 min. Then, 2.5 mL of trichloroacetic acid solution (10%, *w*/*v*) was added into the above mixture, followed by centrifugation at 3000 rpm for 10 min at room temperature, and the supernatant was collected. Afterwards, the supernatant (2.5 mL) was added with 0.5 mL of $FeCl_3$ solution (0.1%, *w*/*v*) and 2.5 mL of ultrapure water. After reacting at room temperature for 10 min, 250 μL of the mixture was added into a 96-well microtiter plate. The absorbance of the sample was measured at 700 nm on a microplate reader. The increase of the absorbance at 700 nm was used to evaluate the reducing ability of the MRPs.

3.3.6. Analysis of Volatile Compounds by GC–MS/SPME

The analysis of volatile compounds in the MRPs was accomplished on a GC-MS instrument (QP2010SE, GK/J-0950, Shimadzu, Excellence in Science, Inc, Tokyo, Japan). A total of 3 μL of the internal standard 1,2-dichlorobenzene (50 μg mL^{-1} in methanol) and 3 mL of the MRPs were added into a 20 mL sealed headspace vial. Then, insert the needle of the SPME sampler into the headspace vial. The sample was equilibrated in the vial at 50 °C for 30 min to extract the volatile substances, while the SPME fiber (75 μm, carboxen/poly-dimethyl siloxane) was suspended above the liquid surface to absorb the extracted volatile substances in each MRP. Afterwards, the adsorbed volatiles were injected into the injection port of a GC-MS and desorbed at 250 °C for 5 min. The instrument was equipped with a DB wax column (30 m × 0.25 mm × 0.25 μm, Agilent Technology, Inc., Folsom, CA, USA) and a mass spectrometer. During the measurement, high-purity helium was used as the carrier gas, and the flow rate was 1.2 mL min^{-1}. The column temperature program was set as follows: the initial temperature was kept at 40 °C for 3 min, increased to 200 °C at 5 °C min^{-1}, and then increased to 230 °C at 10 °C min^{-1}. The scanning range of mass spectrometric detector was in the range of 40–450 m/z.

3.3.7. Descriptive Sensory Analysis of the MRPs

Sensory evaluation of MRPs was performed according to the method of Song et al. [12]. Ten experts with knowledge of flavor evaluation were selected from the Scientific Sensory Evaluation Laboratory of Hefei University of Technology to evaluate the descriptive senses of the MRPs. The evaluation panel consisted of 5 male and 5 female members. Prior to evaluation, unified standards (Table 4) for specific indicators of flavor, including meaty, off-flavor, umami, salty, and overall satisfaction, were set up by full discussion. The evaluation should be performed at room temperature (25 °C), and the sensory evaluation of each sample was repeated three times in parallel.

3.4. Statistical Analysis

The analysis of each sample was repeated triplicates, and the obtained data were analyzed by one-way analysis of variance (ANOVA) with SPSS version 26.0 (SPSS, Inc., Chicago, IL, USA) software. The analyzed data were presented as mean values ± standard deviations (SDs). Significance was considered at ± 5% ($p < 0.05$).

4. Conclusions

In conclusion, we analyzed the effect of ultrasound on the MRPs derived from hydrolyzed soybean meal in an oil-in-water system. The GC-MS analyses of the fatty acid compositions of lard obtained by different treatments showed that ultrasonic treatment not only accelerated the oxidation of lard, but also decreased its unsaturated fatty acid content. The addition of oxidized lard coupling with ultrasound assistance increased the UV absorbance at 294 nm and 420 nm, darkening the color of the obtained MRPs. Due to the effects (mechanical and cavitation effects) of ultrasound, the volatile compounds of the UN-MRPs and UEL-MRPs were significantly increased, compared to the controls of N-MRPs and EL-MRPs, respectively. More importantly, the volatile substances that contributed greatly to the flavor of the MRPs were increased in UN-MRPs and UEL-MRPs, compared to the controls of N-MRPs and EL-MRPs, respectively. The sensory evaluation also showed that ultrasound exerted positive effects on the taste of the obtained MRPs, as the total acceptance of the UN-MRPs and UEL-MRPs was better than that of the N-MRPs and EL-MRPs. Antioxidant tests showed that the UN-MRPs obtained by ultrasound assistance in oil-free system showed better antioxidant activity than the control N-MRPs, while the UEL-MRPs obtained in oil-in water system showed lower antioxidant activity than the control EL-MRPs. The future work will focus on the separation and purification of MRPs to obtain MRPs with different molecular weight ranges and to find out the relationship between the antioxidant activity, the molecular weight of the MRPs, and the effects of ultrasound.

Author Contributions: Conceptualization, X.C., Y.Y. and S.H.; methodology, X.C. and S.D.; validation, X.C., Y.Y. and S.H.; formal analysis, X.C. and S.D.; investigation, X.C., S.D., H.Z. and Y.Y.; resources, X.C., S.H., Z.W. and Y.Y.; data curation, X.C. and S.D.; writing—original draft preparation, X.C. and S.D.; writing—review and editing, X.C., S.H., Z.W. and Y.Y.; visualization, X.C., S.D. and Y.Y.; supervision, X.C. and Z.W.; project administration, X.C.; funding acquisition, X.C., W.H. and Y.Y. All authors have read and agreed to the published version of the manuscript.

Funding: This research was funded by the Anhui Province Science and Technology Major Project (grant numbers 202003a06020018 and 201903b06020007) and Enterprise Entrusted Project (grant number W2022JSKF0112).

Institutional Review Board Statement: Not applicable.

Informed Consent Statement: Not applicable.

Data Availability Statement: The data that support the findings of this study are available from the corresponding author upon reasonable request.

Conflicts of Interest: The authors declare no conflict of interest.

Sample Availability: Not applicable.

References

1. Siewe, F.B.; Kudre, T.G.; Bettadaiah, B.K.; Narayan, B. Effects of Ultrasound-Assisted Heating on Aroma Profile, Peptide Structure, Peptide Molecular Weight, Antioxidant Activities and Sensory Characteristics of Natural Fish Flavouring. *Ultrason. Sonochem.* **2020**, *65*, 105055. [CrossRef] [PubMed]
2. Habinshuti, I.; Chen, X.; Yu, J.; Mukeshimana, O.; Duhoranimana, E.; Karangwa, E.; Muhoza, B.; Zhang, M.; Xia, S.; Zhang, X. Antimicrobial, Antioxidant and Sensory Properties of Maillard Reaction Products (MRPs) Derived from Sunflower, Soybean and Corn Meal Hydrolysates. *LWT* **2019**, *101*, 694–702. [CrossRef]
3. Xia, B.; Ni, Z.J.; Hu, L.T.; Elam, E.; Thakur, K.; Zhang, J.G.; Wei, Z.J. Development of Meat Flavors in Peony Seed-Derived Maillard Reaction Products with the Addition of Chicken Fat Prepared under Different Conditions. *Food Chem.* **2021**, *363*, 130276. [CrossRef] [PubMed]
4. Ye, Y.; Ye, S.; Wanyan, Z.; Ping, H.; Xu, Z.; He, S.; Cao, X.; Chen, X.; Hu, W.; Wei, Z. Producing Beef Flavors in Hydrolyzed Soybean Meal-Based Maillard Reaction Products Participated with Beef Tallow Hydrolysates. *Food Chem.* **2022**, *378*, 132119. [CrossRef] [PubMed]
5. Yu, M.; He, S.; Tang, M.; Zhang, Z.; Zhu, Y.; Sun, H. Antioxidant Activity and Sensory Characteristics of Maillard Reaction Products Derived from Different Peptide Fractions of Soybean Meal Hydrolysate. *Food Chem.* **2018**, *243*, 249–257. [CrossRef]
6. Chen, X.; Jiang, D.; Xu, P.; Geng, Z.; Xiong, G.; Zou, Y. Structural and Antimicrobial Properties of Maillard Reaction Products in Chicken Liver Protein Hydrolysate after Sonication. *Food Chem.* **2021**, *343*, 128417. [CrossRef]
7. Zhang, Y.; Ma, Y.; Ahmed, Z.; Geng, W.; Tang, W.; Liu, Y.; Jin, H.; Jiang, F.; Wang, J.; Wang, Y. Purification and Identification of Kokumi-Enhancing Peptides from Chicken Protein Hydrolysate. *Int. J. Food Sci. Technol.* **2019**, *54*, 2151–2158. [CrossRef]
8. Wei, C.K.; Ni, Z.J.; Thakur, K.; Liao, A.M.; Huang, J.H.; Wei, Z.J. Aromatic Effects of Immobilized Enzymatic Oxidation of Chicken Fat on Flaxseed (*Linum Usitatissimum* L.) Derived Maillard Reaction Products. *Food Chem.* **2020**, *306*, 125560. [CrossRef]
9. Xu, Y.; Chen, Q.; Lei, S.; Wu, P.; Fan, G.; Xu, X.; Pan, S. Effects of Lard on the Formation of Volatiles from the Maillard Reaction of Cysteine with Xylose. *J. Sci. Food Agric.* **2011**, *91*, 2241–2246. [CrossRef]

10. Ma, X.; Zhan, P.; Tian, H.; Wei, Z.; Wang, P. Effects of Different Enzymatic Hydrolyses of Mutton Tallow on the Aroma Characteristics of the Maillard Reaction of Xylose–Cysteine Based on GC-MS, E-Nose, and Statistical Analysis. *Eur. J. Lipid Sci. Technol.* **2020**, *122*, 1900212. [CrossRef]
11. Song, S.; Tang, Q.; Hayat, K.; Karangwa, E.; Zhang, X.; Xiao, Z. Effect of Enzymatic Hydrolysis with Subsequent Mild Thermal Oxidation of Tallow on Precursor Formation and Sensory Profiles of Beef Flavours Assessed by Partial Least Squares Regression. *Meat Sci.* **2014**, *96*, 1191–1200. [CrossRef] [PubMed]
12. Song, S.; Tang, Q.; Fan, L.; Xu, X.; Song, Z.; Hayat, K.; Feng, T.; Wang, Y. Identification of Pork Flavour Precursors from Enzyme-Treated Lard Using Maillard Model System Assessed by GC–MS and Partial Least Squares Regression. *Meat Sci.* **2017**, *124*, 15–24. [CrossRef]
13. Yang, Z.; Xie, J.; Zhang, L.; Du, R.; Cao, C.; Wang, M.; Acree, T.; Sun, B. Aromatic Effect of Fat and Oxidized Fat on a Meat-like Model Reaction System of Cysteine and Glucose. *Flavour Fragr. J.* **2015**, *30*, 320–329. [CrossRef]
14. Yu, H.; Zhong, Q.; Liu, Y.; Guo, Y.; Xie, Y.; Zhou, W.; Yao, W. Recent Advances of Ultrasound-Assisted Maillard Reaction. *Ultrason. Sonochem.* **2020**, *64*, 104844. [CrossRef] [PubMed]
15. Dong, Z.Y.; Liu, W.; Zhou, Y.J.; Ren, H.; Li, M.Y.; Liu, Y. Effects of Ultrasonic Treatment on Maillard Reaction and Product Characteristics of Enzymatic Hydrolysate Derived from Mussel Meat. *J. Food Process Eng.* **2019**, *42*, 1–11. [CrossRef]
16. Yu, H.; Seow, Y.-X.; Ong, P.K.C.; Zhou, W. Effects of High-Intensity Ultrasound and Oil Type on the Maillard Reaction of d-Glucose and Glycine in Oil-in-Water Systems. *npj Sci. Food* **2018**, *2*, 1–2. [CrossRef] [PubMed]
17. Stanic-Vucinic, D.; Prodic, I.; Apostolovic, D.; Nikolic, M.; Cirkovic Velickovic, T. Structure and Antioxidant Activity of β-Lactoglobulin-Glycoconjugates Obtained by High-Intensity-Ultrasound-Induced Maillard Reaction in Aqueous Model Systems under Neutral Conditions. *Food Chem.* **2013**, *138*, 590–599. [CrossRef] [PubMed]
18. Trushenski, J.T.; Rombenso, A.N.; Page, M.; Jirsa, D.; Drawbridge, M. Traditional and Fermented Soybean Meals as Ingredients in Feeds for White Seabass and Yellowtail Jack. *N. Am. J. Aquac.* **2014**, *76*, 312–322. [CrossRef]
19. Li, X.; Zhang, F.; Li, J.; Xia, C.; Li, J. Evaluation Performance of Soybean Meal and Peanut Meal Blends-Based Wood Adhesive. *Polym. Test.* **2022**, *109*, 107543. [CrossRef]
20. Huang, A.; Sun, L.; Lin, F.; Guo, J.; Jiang, J.; Shen, B.; Chen, J. Medical Image Recognition Technology in the Effect of Substituting Soybean Meal for Fish Meal on the Diversity of Intestinal Microflora in Channa Argus. *J. Healthc. Eng.* **2021**, *2021*, 5269169. [CrossRef] [PubMed]
21. Zhang, Z.; Elfalleh, W.; He, S.; Tang, M.; Zhao, J.; Wu, Z.; Wang, J.; Sun, H. Heating and Cysteine Effect on Physicochemical and Flavor Properties of Soybean Peptide Maillard Reaction Products. *Int. J. Biol. Macromol.* **2018**, *120*, 2137–2146. [CrossRef] [PubMed]
22. Yu, H.; Seow, Y.X.; Ong, P.K.C.; Zhou, W. Effects of Ultrasonic Processing and Oil Type on Maillard Reaction of D-Glucose and L-Alanine in Oil-in-Water Systems. *Food Bioprocess Technol.* **2019**, *12*, 325–337. [CrossRef]
23. Chen, K.; Yang, Q.; Hong, H.; Feng, L.; Liu, J.; Luo, Y. Physicochemical and Functional Properties of Maillard Reaction Products Derived from Cod (*Gadus Morhua* L.) Skin Collagen Peptides and Xylose. *Food Chem.* **2020**, *333*, 127489. [CrossRef] [PubMed]
24. Perusko, M.; Al-Hanish, A.; Velickovic, T.C.; Stanic-Vucinic, D. Macromolecular Crowding Conditions Enhance Glycation and Oxidation of Whey Proteins in Ultrasound-Induced Maillard Reaction. *Food Chem.* **2015**, *177*, 248–257. [CrossRef] [PubMed]
25. Abdelhedi, O.; Mora, L.; Jemil, I.; Jridi, M.; Toldrá, F.; Nasri, M.; Nasri, R. Effect of Ultrasound Pretreatment and Maillard Reaction on Structure and Antioxidant Properties of Ultrafiltrated Smooth-Hound Viscera Proteins-Sucrose Conjugates. *Food Chem.* **2017**, *230*, 507–515. [CrossRef] [PubMed]
26. Karangwa, E.; Zhang, X.; Murekatete, N.; Masamba, K.; Raymond, L.V.; Shabbar, A.; Zhang, Y.; Duhoranimana, E.; Muhoza, B.; Song, S. Effect of Substrate Type on Sensory Characteristics and Antioxidant Capacity of Sunflower Maillard Reaction Products. *Eur. Food Res. Technol.* **2015**, *240*, 939–960. [CrossRef]
27. Karangwa, E.; Raymond, L.V.; Huang, M.; Cheserek, M.J.; Hayat, K.; Savio, N.D.; Amédée, M.; Zhang, X. Sensory Attributes and Antioxidant Capacity of Maillard Reaction Products Derived from Xylose, Cysteine and Sunflower Protein Hydrolysate Model System. *Food Res. Int.* **2013**, *54*, 1437–1447. [CrossRef]
28. Zhang, W.; Leong, S.M.; Zhao, F.; Zhao, F.; Yang, T.; Liu, S. Viscozyme L Pretreatment on Palm Kernels Improved the Aroma of Palm Kernel Oil after Kernel Roasting. *Food Res. Int.* **2018**, *107*, 172–181. [CrossRef] [PubMed]
29. Shen, Y.; Hu, L.; Xia, B.; Ni, Z.; Elam, E.; Thakur, K.; Zhang, J.; Wei, Z. Effects of Different Sulfur-Containing Substances on the Structural and Flavor Properties of Defatted Sesame Seed Meal Derived Maillard Reaction Products. *Food Chem.* **2021**, *365*, 130463. [CrossRef] [PubMed]
30. Zhao, J.; Wang, T.; Xie, J.; Xiao, Q.; Cheng, J.; Chen, F.; Wang, S. Formation Mechanism of Aroma Compounds in a Glutathione-Glucose Reaction with Fat or Oxidized Fat. *Food Chem.* **2019**, *270*, 436–444. [CrossRef]
31. Lee, S.E.; Chung, H.; Kim, Y.S. Effects of Enzymatic Modification of Wheat Protein on the Formation of Pyrazines and Other Volatile Components in the Maillard Reaction. *Food Chem.* **2012**, *131*, 1248–1254. [CrossRef]
32. Yu, H.; Seow, Y.X.; Ong, P.K.C.; Zhou, W. Effects of High-Intensity Ultrasound on Maillard Reaction in a Model System of d-Xylose and l-Lysine. *Ultrason. Sonochem.* **2017**, *34*, 154–163. [CrossRef] [PubMed]
33. Tu, Y.; Xu, Y.; Ren, F.; Zhang, H. Characteristics and Antioxidant Activity of Maillard Reaction Products from α-Lactalbumin and 2′-Fucosyllactose. *Food Chem.* **2020**, *316*, 126341. [CrossRef] [PubMed]
34. Han, L.; Li, F.; Yu, Q.; Li, D. In Vitro Antioxidant and Cytoprotective Properties of Maillard Reaction Products from Phloridzin-Amino Acid Model Systems. *J. Sci. Food Agric.* **2018**, *98*, 590–597. [CrossRef] [PubMed]

35. Liu, H.; Jiang, Y.; Guan, H.; Li, F.; Sun-Waterhouse, D.; Chen, Y.; Li, D. Enhancing the Antioxidative Effects of Foods Containing Rutin and α-Amino Acids via the Maillard Reaction: A Model Study Focusing on Rutin-Lysine System. *J. Food Biochem.* **2020**, *44*, 1–12. [CrossRef] [PubMed]
36. Zeng, Q.; Cui, Y.; Su, D.; Bin, T.; Yuan, Y.; He, S. Process Optimization and Anti-Oxidative Activity of Peanut Meal Maillard Reaction Products. *LWT* **2018**, *97*, 573–580. [CrossRef]
37. Li, Q.; Li, X.; Ren, Z.; Wang, R.; Zhang, Y.; Li, J.; Ma, F.; Liu, X. Physicochemical Properties and Antioxidant Activity of Maillard Reaction Products Derived from Dioscorea Opposita Polysaccharides. *LWT* **2021**, *149*, 111833. [CrossRef]

molecules MDPI

Article

Structural Characteristics and Antioxidant Mechanism of Donkey-Hide Gelatin Peptides by Molecular Dynamics Simulation

Rong Liang, Le Xu, Chen Fan, Lele Cao and Xingfeng Guo *

Agricultural Science and Engineering School, Liaocheng University, Liaocheng 252059, China; liangrong8209@163.com (R.L.); xule@lcu.edu.cn (L.X.); fanchen7810@126.com (C.F.); 18846170635@163.com (L.C.)
* Correspondence: guoxingfeng@lcu.edu.cn

Abstract: This study aimed to explore the structural characteristics and antioxidant mechanism of donkey-hide gelatin peptides. After hydrolysis and ultrafiltration treatment, five gelatin peptides with different molecular weights (MWs) were obtained. Amino acid analysis showed that gelatin peptides with different MWs contained a large number of amino acids, including G, P, E, N, A, and R, and differences were noted in the content of various amino acids. Fourier transform infrared spectroscopy and circular dichroism revealed that these gelatin peptides differed in terms of the peak strength of functional groups and number of secondary structures. Moreover, 26 pentapeptides/hexapeptides were identified. Among them, we investigated by molecular docking how PGPAP, which has the best antioxidant activity, may interact with the Keap1 protein. The results showed that the PGPAP-Keap1 complex had a stable conformation, and Arg415, Gly462, Phe478, and Tyr572 were the key residues involved in the binding of the peptide PGPAP to Keap1. Our results demonstrated that PGPAP could serve as a bioactive peptide with antioxidant activity.

Keywords: gelatin peptides; antioxidant activity; structural characteristics; Keap1 protein; molecular dynamics simulations

Citation: Liang, R.; Xu, L.; Fan, C.; Cao, L.; Guo, X. Structural Characteristics and Antioxidant Mechanism of Donkey-Hide Gelatin Peptides by Molecular Dynamics Simulation. *Molecules* **2023**, *28*, 7975. https://doi.org/10.3390/molecules28247975

Academic Editors: Shudong He, Wei Xu and Zengwang Guo

Received: 20 October 2023
Revised: 30 November 2023
Accepted: 1 December 2023
Published: 6 December 2023

1. Introduction

Free radicals in the body can react with biologically active molecules to cause cell or tissue damage. Oxidative stress is linked to many diseases, such as inflammation and depression. Some synthetic antioxidants can scavenge free radicals and prevent food from spoiling. Although synthetic antioxidants possess strong antioxidant activity, they may cause potential harm and toxicity, which limit application in food products [1]. Therefore, the development of healthy and natural antioxidants has become a research trend. Some recent studies have demonstrated that many animal and plant protein hydrolysates have good antioxidant activities [2]. In general, an active antioxidant peptide is composed of 3–16 amino acid residues. Antioxidant peptides have the ability to scavenge free radicals, inhibit lipid peroxidation, supply hydrogen, and chelate metal ions. Considering that foodborne antioxidant peptides are safe and nontoxic, research on antioxidant peptides has gradually become a hot topic.

Gelatin is a colloidal protein extracted from the collagen of animal bones, hides, tendons, and other connective tissues. It is a straight chain polymer formed by crosslinking 18 amino acids and peptides [3]. Currently, gelatin is used as a frozen food improver, confectionery additive, beverage clarifying agent, meat product improver, dairy product additive, food coating material, and sugar coating agent in the food industry [4]. Peptides derived from gelatin possess strong antioxidant properties. For example, the EC50 values of DPPH, OH, and O^{2-} free radicals cleared by the skin gelatin hydrolysates of bluefin leatherjacket (*Navodon septentrionalis*) were 5.227, 1.147, and 4.752 mg/mL, respectively [5].

Moreover, after enzymatic hydrolysis via Protamex, these hydrolysates could scavenge DPPH free radicals and chelate Fe^{2+} ions, confirming their antioxidant effects [6].

Extensive research has proven that the antioxidant activity of peptides is affected by various physicochemical properties, including the molecular weight (MW) of peptides as well as the type and sequence of amino acids [7]. Zhang et al. revealed that the MW distribution of peanut polypeptides with the best antioxidant activity was mainly observed below 1400 Da [8]. Liu et al. obtained whey protein antioxidant peptides with different MWs and found that the peptide group with the smallest MW (<10 kDa) had the strongest antioxidant activity [9]. Sun et al. hydrolyzed porcine hemoglobin using pepsin and obtained four groups of antioxidant peptides; they revealed that that the peptide group with a MW of <3 kDa had the strongest superoxide anion scavenging ability and lipid peroxidation inhibition ability [10]. In addition, amino acid composition is a factor affecting the bioactivity of antioxidant polypeptide molecules. The enzymatic hydrolysate of smooth hound (*Mustelus mustelus*) muscle protein with strong antioxidant activity contained a high proportion of highly active amino acids, including Leu, His, Tyr, and Met [11]. Moreover, antioxidant peptides require a certain spatial structure to exert their antioxidant effects. According to a previous study, the activity of antioxidant peptides with a certain spatial structure was higher than that of amino acid mixtures with the same mass ratio [12]. However, the relationship between the bioactivity and structure of antioxidant peptides as well as the interaction between the active sites of peptide chain and amino acid residues remain unclear. Exploring the structure–activity relationship of antioxidant peptides not only helps predict the antioxidant activity of polypeptides but also provides a theoretical reference for subsequent efficient screening and synthesis of highly bioactive peptides.

Another important aspect in studying antioxidant peptides is to reveal the antioxidant mechanism. In particular, molecular docking has gradually become an important approach to screen peptides and explore their antioxidant mechanism. It is well-known that the Keap1-Nrf2 pathway is important for human cells to cope with oxidative stress. Under oxidative stress, the body can activate this pathway to promote the expression of a series of endogenous antioxidant proteins, thereby improving the body's antioxidant capacity to resist oxidative stress [13]. Many scholars have used this pathway combined with molecular docking to explore the peptides' antioxidant mechanism. Li et al. screened 20 small peptides with potential antioxidant activity using molecular docking, among which DKK and DDW showed the strongest ability to dock and bind to Keap1 proteins [14]. Agrawal et al. successfully isolated and purified two peptides from millet protein hydrolysates, namely, TSSSLNMAVRGGLTR and STTVGLGISMRSASVR [15]. Molecular docking studies have shown that interactions between Ser/Thr residues and free radicals are the main source of antioxidant activity of the two peptides. Tonolo et al. extracted and identified 23 peptides from fermented milk protein, among which NTVPAKSCQAQPTTM, QGPIVLNPWDQVKR, and APSFSDIPNPIGSENSE showed a strong ability to dock and bind to Keap1 [16]. These three peptides have certain inhibitory effects on Keap1-Nrf2 PPI; moreover, they can upregulate the expression level and activity of antioxidant enzymes by activating the Keap1-Nrf2 pathway [17]. These studies provide a significant reference for exploring the antioxidant mechanism of active peptides.

This study aimed to explore the structure–activity relationship of donkey-hide gelatin peptides using amino acid analysis, Fourier transform infrared (FTIR) spectroscopy, and circular dichroism (CD). In addition, gelatin peptides were identified, and the antioxidant peptides with a known sequence were synthesized. The antioxidant activity of gelatin peptides was analyzed, and the most active peptides were docked to the Keap1 protein through molecular docking. Moreover, the stability of the peptide–Keap1 complex was verified using molecular dynamic (MD) simulations to explore its antioxidant mechanism. This research may indicate the molecular mechanism of active peptides and provide a reference for application of gelatin peptides.

2. Results and Discussion

2.1. Antioxidant Activity of Gelatin Peptides with Different MWs

After enzymatic hydrolysis and ultrafiltration of donkey-hide gelatin, five MW groups of gelatin polypeptides were obtained, including those with a MW of >30, 10–30, 3–10, 1–3, and <1 kDa. Then, the antioxidant activity of gelatin polypeptides was tested. The results are shown in Figure 1a. Peptides with an MW of 1–3 kDa had the highest antioxidant activity, with DPPH and ABTS free radical scavenging abilities of 32.33% ± 2.19% and 94.13% ± 0.37%, respectively. A previous study reported that the antioxidant activity of peptides with a certain spatial structure was higher than that of amino acid mixtures with the same mass ratio [12], and the smaller the MW of peptides, the higher the antioxidant activity. However, in this study, peptides with the smallest MW (<1 kDa) did not show the highest antioxidant activity, which may be due to the presence of more free amino acids in their system. Ngoh et al. also obtained consistent results by separating and purifying pinto protein-hydrolyzed products; in other words, the peptide group with a MW of <3 kDa had the strongest antioxidant activity [18].

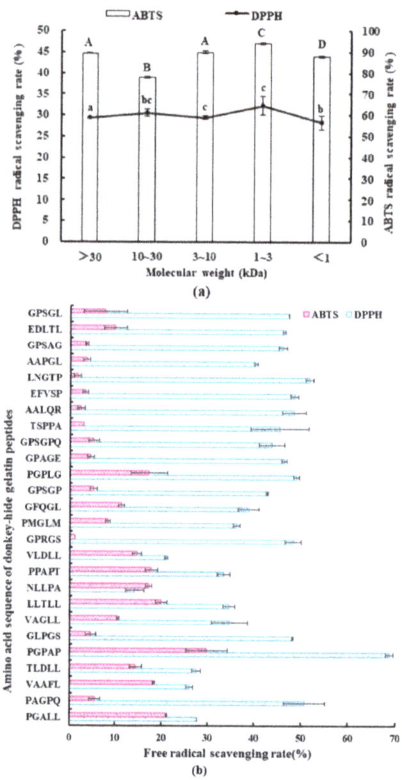

Figure 1. Antioxidant activity of donkey-hide gelatin peptides. (**a**) DPPH and ABTS scavenging ability of donkey-hide gelatin peptides with different molecular weights. The same capital letters A–D mean that the variance of ABTS scavenging ability between two samples is not significant ($p > 0.05$), and the different letters A–D mean significant ($p < 0.05$); the same small letters a–d mean that the variance of DPPH scavenging ability between two samples is not significant ($p > 0.05$), and the different letters a–d mean significant ($p < 0.05$). (**b**) DPPH and ABTS scavenging ability of donkey-hide gelatin peptides with different amino acid sequences.

2.2. Amino Acid Composition of Gelatin Peptides with Different MWs

The antioxidant activity of peptides is closely related to their structural properties, such as MW, amino acid composition, and peptide configuration. To explore the structure–activity relationship, the amino acid composition, functional groups, and secondary structure of polypeptides with different MWs were detected and analyzed. The amino acid composition (a total of 17 amino acids) of each polypeptide group was analyzed (Table 1). Many amino acids, such as G, P, E, N, A, and R, were identified in each group of peptide hydrolysates. In addition, each group contained a certain amount of M, C, and H. The amino acid composition of a peptide has a crucial effect on its antioxidant activity, and some amino acids may be detected in the active sites of antioxidant peptides [19]. A previous study revealed that the E-L sequence contributed significantly to the free radical scavenging ability of the short antioxidant peptide YFYPEL [20,21]. Chi et al. (2015) reported that YLMSR and VLYEE had higher antioxidant activity than MILMR, which might be related to the presence of Y [22]. Saito et al. revealed that peptides containing Y had stronger antioxidant activity than those containing H; moreover, peptides containing W or Y at the carboxyl terminal showed stronger free radical scavenging ability [23]. In this study, although the amino acid types were identical among the peptide groups, differences were noted in the contents of various amino acids. For example, the contents of most of the amino acids (e.g., N and E) in the group with a MW of >30 kDa were higher because the overall amino acid content was higher in this group of large proteins, which are not completely hydrolyzed. However, the contents of G and I were the highest in peptides with a MW of 1–3 kDa. The difference in amino acid composition among different groups may be responsible for their different antioxidant activities.

Table 1. (a). Amino acid composition of donkey-hide gelatin peptides with different molecular weights. (b). Amino acid sequence identification of donkey-hide gelatin peptides (MW < 1 kDa).

(a)

No.	Amino Acid Type	Amino Acid Content of Donkey-Hide Gelatin Peptides (mg/g)				
		>30 kDa	10–30 kDa	3–10 kDa	1–3 kDa	<1 kDa
1	Asn (N)	73.189	63.137	61.07	60.278	50.186
2	Thr (T)	29.513	25.766	27.857	26.285	23.272
3	Ser (S)	39.896	34.747	34.871	35.864	33.833
4	Glu (E)	107.354	90.651	86.697	88.686	78.885
5	Gly (G)	128.919	118.312	124.697	129.384	115.45
6	Ala (A)	63.607	60.372	56.137	59.293	59.261
7	Cys (C)	11.416	11.579	15.318	14.41	9.948
8	Val (V)	25.303	23.089	21.879	22.12	21.671
9	Met (M)	23.675	22.583	20.419	21.044	20.839
10	Ile (I)	32.001	31.48	33.731	35.121	29.606
11	Leu (L)	36.751	33.759	32.873	35.278	32.613
12	Tyr (Y)	16.381	13.858	14.759	15.526	12.439
13	Phe (F)	28.169	26.427	26.334	25.897	24.547
14	His (H)	10.438	9.922	9.249	9.64	8.856
15	Lys (K)	43.891	41.852	39.377	39.551	39.326
16	Arg (R)	60.535	55.137	44.898	51.678	50.074
17	Pro (P)	113.935	96.871	92.763	95.652	96.794

(b)

No.	Amino Acid Sequence	Leading Razor Protein	Protein Names	Gene Names	Score
1	GPSGL	A0A8C4MG85	Obscurin, cytoskeletal calmodulin and titin-interacting RhoGEF	OBSCN	61.28
2	EDLTL				57.71
3	GPSAG				52.86
4	AAPGL				25.12
5	LNGTP				25.08
6	EFVSP				24.47
7	AALQR				24.30
8	TSPPA				20.54

Table 1. *Cont.*

(b)					
No.	**Amino Acid Sequence**	**Leading Razor Protein**	**Protein Names**	**Gene Names**	**Score**
9	GPSGPQ	B9VR89	Collagen alpha-2 type I chain	COL1A2	162.47
10	GPAGE				63.06
11	PGPLG				55.36
12	GPSGP				50.40
13	GFQGL				32.63
14	PMGLM				26.57
15	GPRGS				22.12
16	VLDLL	A0A8C4MF50	Ubiquitin protein ligase E3 component n-recognin 4	UBR4	43.99
17	PPAPT				38.19
18	NLLPA				33.54
19	LLTLL				23.49
20	VAGLL				22.99
21	GLPGS	A0A8C4L3F1	SZT2 subunit of KICSTOR complex	SZT2	56.75
22	PGPAP				61.37
23	TLDLL	A0A8C4LPB4	Neurobeachin like 2	NBEAL2	45.11
24	VAAFL				28.74
25	PAGPQ	A0A8C4KVK0	Dynein heavy chain domain 1	DNHD1	81.36
26	PGALL	A0A8C4MZE5	Zinc finger homeobox 4	ZFHX4	25.12

2.3. Functional Groups and Secondary Structures of Gelatin Peptides with Different MWs

As shown in Figure 2, all gelatin peptides with different MWs showed absorbance in N–H (3500–3300 cm^{-1}), O–H (3300–2500 cm^{-1}), unsaturated C–H (>3000 cm^{-1}), saturated C–H (3000–2800 cm^{-1}), –C=O (1850–1600 cm^{-1}), amide (1700–1600 cm^{-1}), C=C (1680–1620 cm^{-1}), C–N (1360–1180 cm^{-1}), C–O (1300–1080 cm^{-1}), S=O (1220–1040 cm^{-1}), C–O–C (1150–900 cm^{-1}), and $-NO_2$ (1600–1500 cm^{-1}) bands. Some of these bonds were found in amino acids, while some do not exist in amino acids such as S=O and $-NO_2$, which may produce absorption peaks due to stretching between amino acids. Gelatin peptides possessed the same absorption bands because they came from the same source and had the same amino acid composition. However, the peak strength of functional groups varied in peptide groups with different MWs (Figure 2a). This finding may be related to the different amino acid contents, indicating that the spatial structure of the polypeptide aqueous solution is different. It is certain that antioxidant peptides require a spatial structure to exert their antioxidant effects. Differences in spatial structure can affect the exposure of certain active groups and thus affect the antioxidant activity of polypeptides [12].

All CD spectra of gelatin peptides with different MWs showed a negative peak at approximately 200 nm, which is considered a p–p band (Figure 3a). In addition, a small positive band was observed at approximately 225 nm. Based on the CD spectra, polyproline II helix and random coil conformations were detected, with different ellipticity values at wavelengths below 220 nm indicating different content of conformations [24]. As shown in Figure 3b, gelatin peptides presented a high content of antiparallel conformations, followed by beta-turn, whereas random coil, parallel, and alpha-helix conformations were relatively rare. Although gelatin peptides with different MWs showed similar trends in the content of different secondary structures, some differences were also noted. These subtle differences can affect the antioxidant activity. For example, a small change in random coil conformation can influence the antioxidant activity [25]. Antioxidant peptides require a certain spatial structure to exert their antioxidant effects. The best free radical scavenging ability can be achieved when some special groups associated with antioxidant activity are fully exposed [26]. There may be some oligopeptides in gelatin peptides solution, which could not form secondary conformations alone, but some oligopeptides with a very short length may exist in the form of peptide complexes to exhibit a secondary structure [27]. Therefore, differences in the secondary structure of gelatin peptides is another factor affecting antioxidant activities.

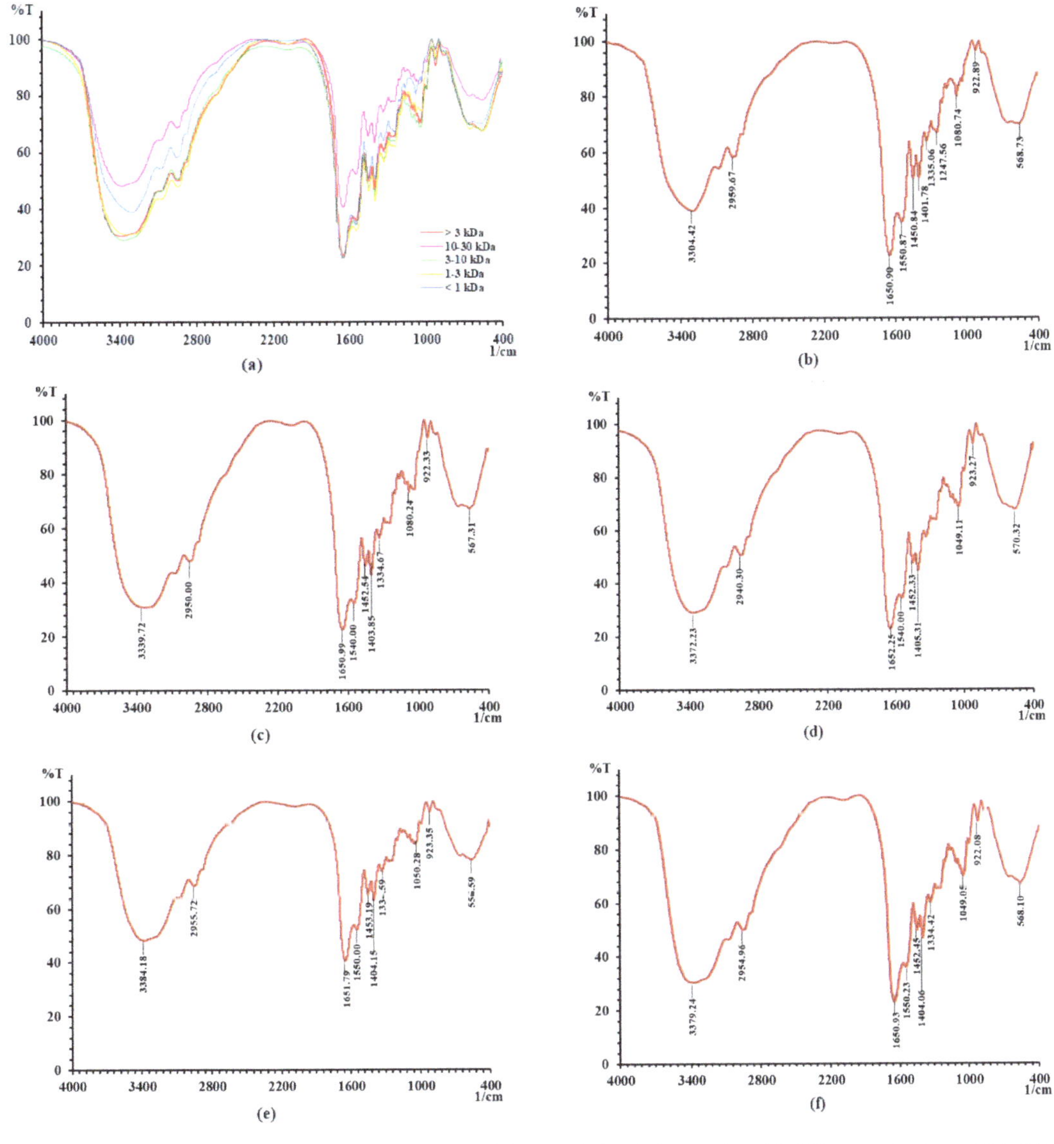

Figure 2. The FTIR spectra of donkey−hide gelatin peptides with different molecular weights. (**a**) FTIR spectra of all donkey−hide gelatin peptides, (**b**) FTIR spectra of donkey-hide gelatin peptide with MW of <1 kDa, (**c**) FTIR spectra of donkey-hide gelatin peptide with MW of 1–3 kDa, (**d**) FTIR spectra of donkey-hide gelatin peptide with MW of 3–10 kDa, (**e**) FTIR spectra of donkey-hide gelatin peptide with MW of 10–30 kDa, and (**f**) FTIR spectra of donkey−hide gelatin peptide with MW of >3 kDa.

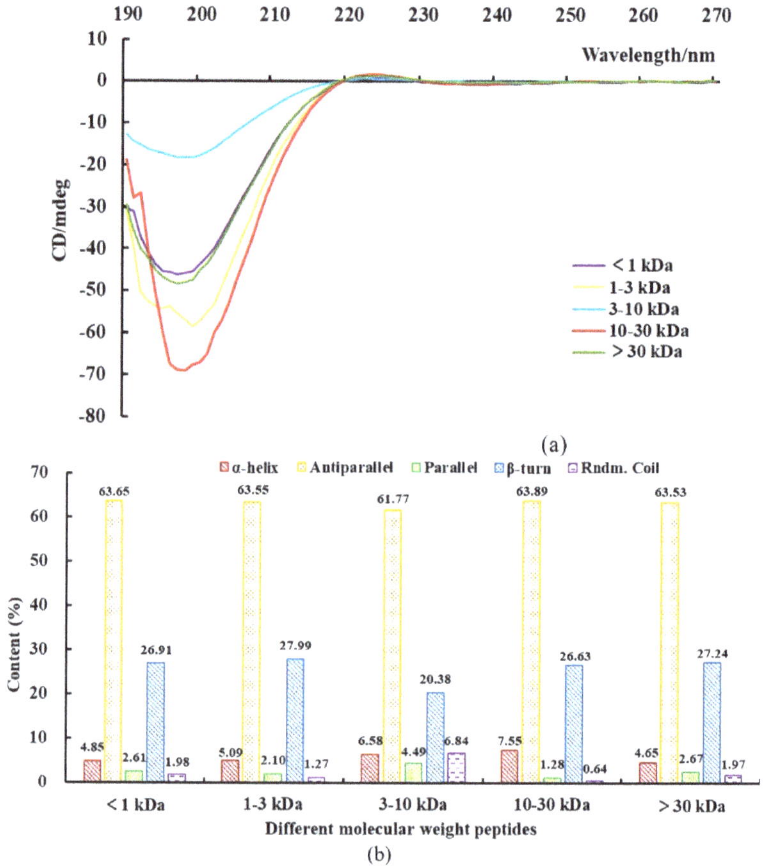

Figure 3. Secondary structure results of donkey–hide gelatin peptides with different molecular weights. (**a**) Circular dichroism spectrum, and (**b**) secondary structure content.

2.4. Sequence Identification and Antioxidant Activity of Synthetic Peptides

To identify the key sequence responsible for antioxidant activity in gelatin polypeptides, LC–MS/MS analysis of the polypeptide group with a MW of 1–3 kDa was performed, and 26 pentapeptides/hexapeptides were identified (Table 1b). These peptides were derived from seven different proteins, including eight peptides from obscurin, cytoskeletal almodulin, and titin-interacting RhoGEF; seven from collagen alpha-2 type I chain; five from ubiquitin protein ligase E3 component n-recognin 4; two from the SZT2 subunit of the KICSTOR complex; two from neurobeachin-like 2; one from dynein heavy chain domain 1; and one from zinc finger homeobox 4. In addition, the higher the overall score of peptide identification, the more reliable the identification results. The identified peptides were synthesized chemically, and their antioxidant activities were analyzed via DPPH and ABTS scavenging ability assays (Figure 1b). The top five peptides showing DPPH scavenging ability included PGPAP, LNGTP, PAGPQ, PGPLG, and EFVSP, whereas those showing ABTS scavenging ability included PGPAP, PGALL, LLTLL, VAAFL, and PPAPT. PGPAP showed the best antioxidant activity, with DPPH and ABTS scavenging abilities of 68.83% ± 1.84% and 29.82% ± 4.51%, respectively. Previous research reported that the antioxidant activity of Leu-Leu-Pro-His-His (soy protein peptide) decreased after the removal of C-terminal His, whereas no effect was observed on antioxidant activity when the N-terminal Leu was removed [28]. Moreover, His and Pro play an important role in the antioxidant activity sequence because the peptide Pro-His-His has high antioxidant activity, and Tyr does not enhance the activity when introduced at the Pro or His location. In addition, the N-terminal of the maize antioxidant peptide contains the hydrophobic amino acid Leu, which enhances the interaction between antioxidant peptide and fatty acid and improves its ability to trap lipid free radicals [29]. In this study, most peptides with higher antioxidant activities contained Pro-like PGPAP and PAGPQ. Some peptides contained the hydrophobic amino acid Leu at the N-terminal, such as LNGTP and LLTLL, with high ABTS scavenging ability.

2.5. Molecular Dynamics (MD) Simulations and Stability Analysis of PGPAP Binding to Keap1

The PGPAP, which showed the best antioxidant activity, was docked onto Keap1 protein, to explore the antioxidant mechanism of gelatin peptides. Using AutoDock, the predicted binding energy of PGPAP to the Keap1 protein was determined to be −5.375 kcal/mol. Then, stability of PGPAP binding to Keap1 was characterized by MD simulations of 50 ns length. Root-mean-square deviation (RMSD) is an important parameter for measuring the stability of the system. Figure 4a shows the changes in the RMSD values of all atoms in the PGPAP–Keap1 complex system over time. The RMSD of the complex increased significantly at the initial stage of simulation. Subsequently, the RMSD of the system began to stabilize at approximately 20 ns, and the average RMSD of the whole process was 0.389 ± 0.086 nm. Therefore, the dynamic simulation based on RMSD was stable and reliable, which could be used for further analysis.

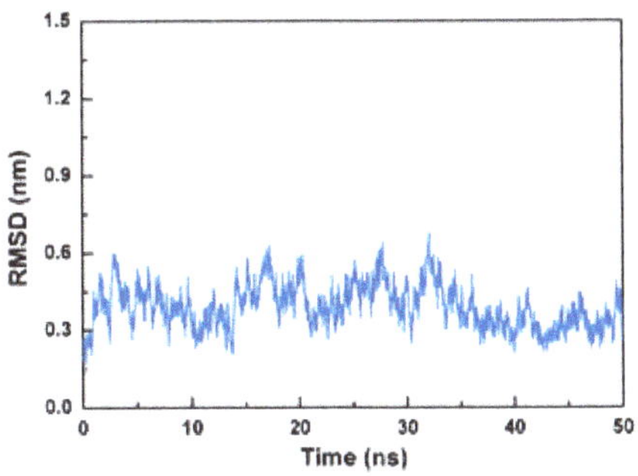

(a)

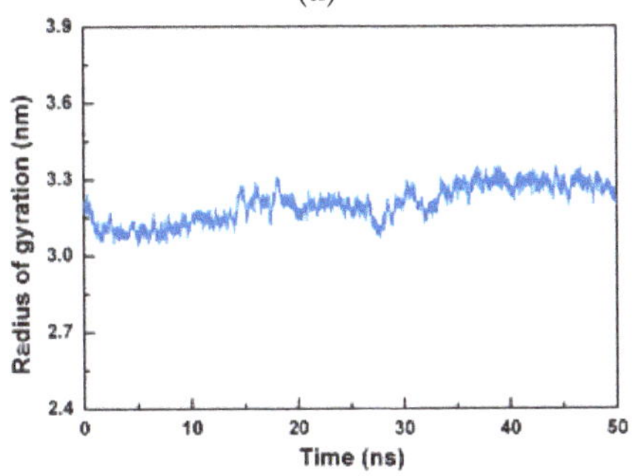

(b)

Figure 4. *Cont.*

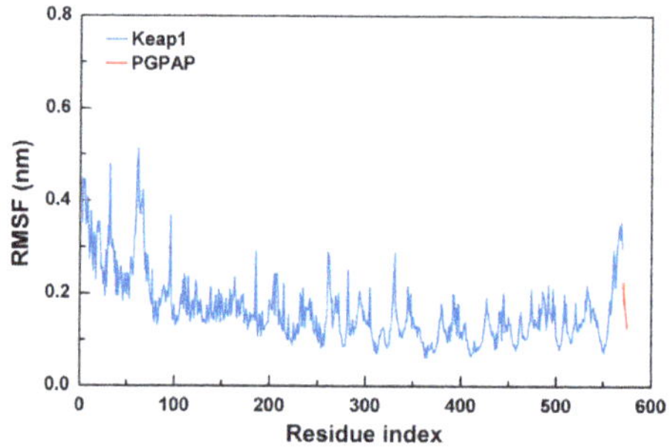

(**c**)

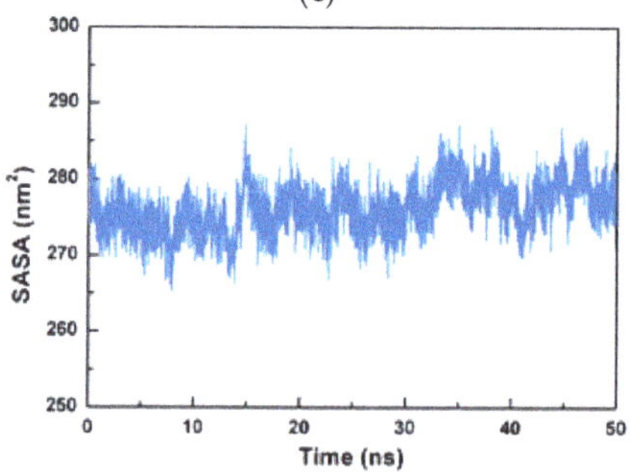

(**d**)

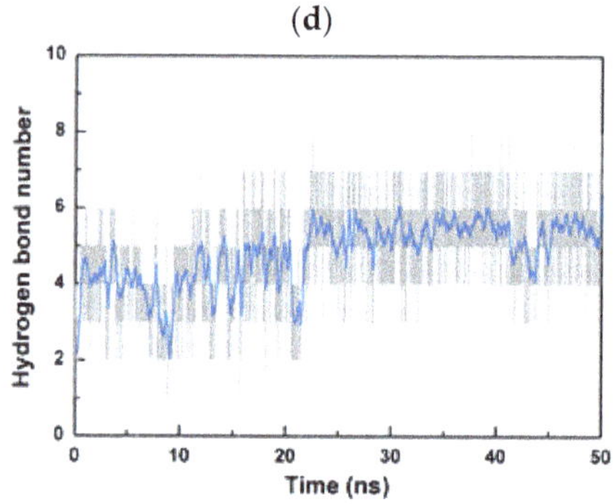

(**e**)

Figure 4. Root-mean-square deviation (RMSD), radius of gyration (Rg), root-mean-square fluctuation (RMSF), solvent-accessibility surface area (SASA) and hydrogen bond number curves of the protein backbone (Cα) atoms. (**a**) RMSD curves of the PGPAP-Keap1 complex. (**b**) Rg curves of the PGPAP-Keap1 complex. (**c**) RMSF curves of the PGPAP-Keap1 complex. (**d**) SASA curves of the PGPAP-Keap1 complex. (**e**) Hydrogen bond number curves of the PGPAP-Keap1 complex.

Radius of gyration (Rg) indicates the compactness of the protein's overall structure [30]. If protein folding is stable, then Rg can maintain a relatively stable value, and a large change in Rg indicates that the system is more expansive. As shown in Figure 4b, the Rg of the whole system decreased at the beginning of the simulation and then gradually increased. The whole complex interacts with solvent molecules, leading to a certain expansion of the protein structure, followed by complete stabilization at approximately 30 ns. The Rg value of the PGPAP–Keap1 complex during simulation reached approximately 3.202 $\pm$ 0.072 nm. This finding indicated that the structure of the PGPAP–Keap1 complex was compact and stable.

Root-mean-square fluctuation (RMSF) indicates the flexibility of amino acid residues in a protein. As shown in Figure 4c, the RMSF value of PGPAP indicated that the whole protein structure was stable during simulation and the polypeptide molecule did not alter greatly after binding to proteins. Thus, it was determined that such a molecule could stably bind to the active pocket of the Keap1 protein.

The solvent-accessibility surface area (SASA) was used to characterize the molecular surface area of the protein system, which may be in contact with the solvent. Consequently, the structural volume changes of proteins in a solvent were analyzed. A high SASA value indicates a loose protein structure and large volume. The SASA value of the PGPAP–Keap1 complex did not increase or decrease significantly during simulation, and the mean value of SASA was 276.219 ± 3.312 nm^2 during the whole process (Figure 4d). The small fluctuation indicated that the complex formed by the peptide and Keap1 protein remained stable throughout the dynamic simulation. Therefore, PGPAP could tightly bind to Keap1 in stable conformations, thereby confirming its potential application as an antioxidant peptide.

2.6. Molecular Mechanism Analysis of Peptide PGPAP Binding to Keap1

Variations in the number of hydrogen bonds between the Keap1 protein and PGPAP polypeptide were analyzed (Figure 4e). The number of hydrogen bonds between them decreased slightly when the whole PGPAP–Keap1 complex underwent conformational change and stabilized, and the number of hydrogen bonds increased further. The average number of hydrogen bonds between Keap1 and PGPAP was 5.36 when the RMSD value was stable at 25–50 ns. Consequently, PGPAP polypeptide molecules could stably bind to Keap1 protein.

This study extracted the complex conformation of polypeptides after MD simulation and analyzed the interaction among them to confirm the main interaction regions between Keap1 and PGPAP as well as key amino acids involved in the interaction. As shown in Figure 5a, the polypeptide molecules are primarily bound to the hydrophobic cavities of the Keap1 protein. Based on the analysis shown in Figure 5b,c, the O atom (carboxyl oxygen from amide bond of the peptide PGPAP), which acts as the hydrogen bond acceptor formed a hydrogen bond with the surrounding amino acids like Arg415 and Gly462. Meanwhile, the polypeptide molecule also formed a π–π stacking interaction with the benzene ring on the surrounding amino acids—Phe478 and Tyr572—which further enhanced the affinity of the polypeptide molecule towards the Keap1 protein. The results indicated that the peptide PGPAP can stably bind to the active cavity of the Keap1 protein to play the corresponding biological role. In addition, the hydrophobic interfaces between PGPAP and Keap1 are shown in Figure 5d. The peptide PGPAP was bent and folded into the hydrophobic pocket of Keap1. The residues Gly462 and Phe478 were responsible for forming the PGPAP-induced hydrophobic interface, which helped the polypeptide molecules enter the active pocket and stably bind to the Keap1 protein.

The human Keap1 protein has five domains, namely, N-terminal region, POZ region (also known as the BTB region), IVR region, KELCH region, and C-terminal region [31]. The human Nrf2 protein can be divided into six conserved domains, namely, Neh1, Neh2, Neh3, Neh4, Neh5, and Neh6 [32]. The Neh2 region is responsible for binding to the Keap1 protein. Under normal physiological conditions, Nrf2 is continuously produced in cells. Keap1 binds to cytoplasmic actin through its POZ region to form homologous dimers; moreover, it binds to the DLG and ETGE sequences of the Neh2 region of Nrf2 protein through two Klech regions of the dimer [33]. In addition, the Keap1 protein dimer binds to Cul3 through its POZ region, which may promote the ubiquitination of the Nrf2 bound to the Keap1 protein dimer, thereby leading to the degradation of ubiquitinated Nrf2 and ensuring that free Nrf2 proteins in the body are present at low levels [34]. When the cells are under oxidative stress, foreign oxidants can alter the conformation of key cysteine residues in POZ and IVR regions of Keap1, thereby altering the conformation of the Keap1 dimer. This conformational change may interfere with the Keap1–Nrf2 interaction. If foreign antioxidants affect the Keap1–Nrf2 interaction, which may increase the intracellular Nrf2 content, then such substances can activate the pathway and improve the antioxidant capacity of the body. In this study, MD simulation suggested that PGPAP could stably bind to Keap1, indicating that it has good potential antioxidant function. Further experiments are warranted to explore its intracellular antioxidant activity and mechanism using HepG2 cell models.

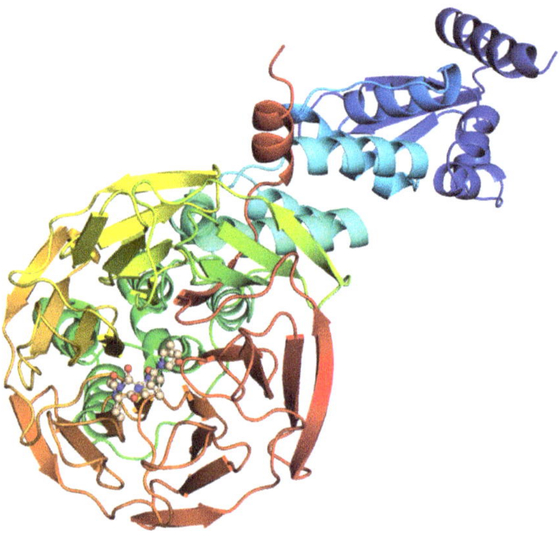

(**a**)

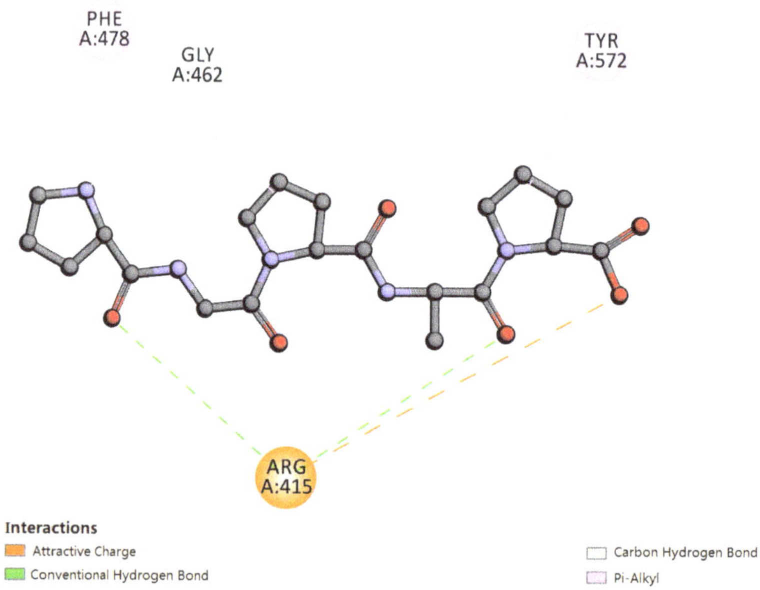

(**b**)

Figure 5. *Cont.*

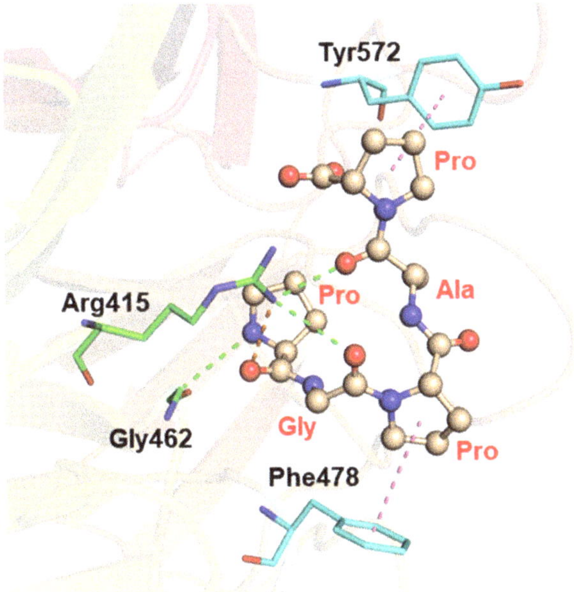

(c)

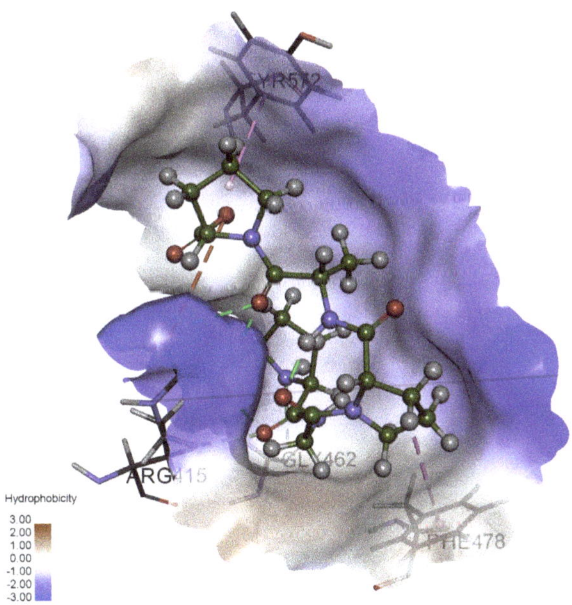

(d)

Figure 5. Docking interactions of PGPAP with Keap1. (**a**) Crystal structure of Keap1 bound with PGPAP, (**b**) 2D diagram how PGPAP interacts with Keap1. The dotted green lines represent hydrogen bonding, orange represent salt bridge interactions, and pink represent π–π stacking interaction. (**c**) 3D diagram how PGPAP interacts with Keap1, and (**d**) hydrophobic interfaces of PGPAP binding to Keap1. Blue and gray colors represent the hydrophilic and hydrophobic parts of the protein surface, respectively.

3. Materials and Methods

3.1. Materials and Reagents

Donkey-hide gelatin (batch number: 2110014) was purchased from Dong-E-E-Jiao Co., Ltd. (Liaocheng, China). Alcalase 2.4L was obtained from Novozymes Biotech. Co., Ltd. (CPH, Denmark). Potassium bromide (KBr), potassium persulfate ($K_2S_2O_8$), formic acid, acetonitrile, 2, 2-diphenyl-1-picrylhydrazyl (DPPH), 2, 2-azino-bis (3-ethylbenzothiazoline-6-sulfonic acid) diammonium salt (ABTS), and amino acid standard (AAS18) were obtained from Sigma Chemicals Co. (USA). All other chemical reagents were of analytical grade purity and obtained from Peking Chemical Plant (Beijing, China).

3.2. Preparation of Gelatin Peptides with Different MWs

Gelatin peptides were prepared according to the method of Yu et al. [35]. Briefly, 5 g of donkey-hide gelatin was dissolved in 120 mL of water and then simmered in a water bath at 80 °C for 30 min. After cooling at 63 °C, the pH of the donkey-hide gelatin solution was adjusted to 10.5 using NaOH (1 mol/L), and Alcalase 2.4L (enzyme to substrate ratio of 9%) was added to initiate hydrolysis. During enzymolysis, the pH range was maintained within ±0.05. After hydrolysis for 3 h, the enzyme was inactivated in a water bath at 90 °C for 10 min, followed by centrifugation of gelatin hydrolysates for 10 min at 10^4 r/min and 4 °C. Then, the supernatant was filtered through Amicon ultra centrifugal filters (Millipore, Burlington, MA, USA) with interceptions of 30, 10, 3, and 1 kDa. Finally, the gelatin-derived peptide solutions with different MWs were obtained.

3.3. Analysis of the Amino Acid Composition of Gelatin Peptides

Overall, 100 mg of gelatin peptide powder was placed in a hydrolytic tube, and 10 mL of HCl (6 M) was added to the tube in a ratio of 1:1. Nitrogen was blown into the tube for 30 s, after which the tube was sealed. Then, the sample was hydrolyzed in an oil bath at 110 °C for 24 h. After hydrolysis, it was cooled to room temperature and filtered through a 0.45-μm membrane into a 50-mL volumetric bottle. Next, the samples were deacidified and filtered through a 0.45-μm filter for machine analysis using Biochrom 30+ amino acid analyzer (Biochrom Ltd., FCE, Cambridge, UK). An Na-type cationic resin chromatography column (200 mm × 4.6 mm) was used for gradient elution. The temperature of the separation column was 55 °C–65 °C–77 °C, and the temperature of the reaction tank was 138 °C. The flow rate of the buffer solution was 20 mL/h, whereas that of the reaction solution was 10 mL/h. The sample volume was 50 μL, and the ultraviolet (UV) detection wavelengths were 570 nm and 440 nm.

3.4. Analysis of the Functional Groups of Gelatin Peptides via FTIR Spectroscopy

FTIR spectroscopy was performed with a resolution of 4 cm^{-1} over the range of 4000–400 cm^{-1} using a UV spectrometer (Thermo Nicolet iS5, Thermo Fisher, Waltham, MA, USA). Briefly, 2 mg of peptide powder and 200 mg of dried KBr were mixed to prepare the sample pellets. The pellets were placed in a sample compartment and measured 32 times at a scanning speed of 2.8 mm/s. After background correction (dried KBr spectrum used as background), the sample spectra were obtained.

3.5. Analysis of the Secondary Structure of Gelatin Peptides via CD

The CD spectra and secondary structures were analyzed using Chirascan 100 Circular Dichroism Spectropolarimeter (Applied Photophysics Ltd., FCE, Surrey, UK). Before the detection test, the nitrogen flow rate was adjusted to between 0.15 and 0.2 MPa; nitrogen was passed for 30 min, and the gas in the optical path of the instrument was removed. The gelatin peptide powder was mixed with deionized water to form a liquid sample (1 mg/mL), which was then slowly poured into the sample pool with an optical path of 1.0 cm for the test. The wavelength range was set from 190 to 270 nm; optimal bandwidth was set to 1 nm; scanning speed was set to 100 nm/min; and the scan time was set to three accumulations. Finally, the data were analyzed using the following equation [36]:

$$[\theta] = \frac{\theta}{10LCN}$$

where θ indicates instrument readings, L indicates the path length (cm), C indicates concentration of the peptide (mol/L), and N indicates the number of amino acid residues in the peptide.

3.6. Determination of the Antioxidant Activity of Gelatin Peptides

3.6.1. DPPH Radical Scavenging Assay of Gelatin Peptides

The reaction system comprised of 96-well microplates containing 100 μL of gelatin peptide solution (6 mg/mL), 100 μL of fresh DPPH (0.6 mM), and 100 μL of methanol. Subsequently, samples were shaken vigorously and incubated at 25 °C for 30 min under dark conditions. Then, the absorbance was measured at a wavelength of 515 nm (using methanol as a blank). Finally, the DPPH radical scavenging activity of gelatin peptides was estimated using the following equation:

$$\text{DPPH radical scavening activity } (\%) = (1 - \frac{Asample}{Ablank}) \times 100$$

3.6.2. ABTS Radical Scavenging Assay of Gelatin Peptides

The ABTS radical scavenging assay was performed in accordance with the method of Kozics et al. with minor modifications [37]. Overall, 10 mL of ABTS (7 mM) and 10 mL of $K_2S_2O_8$ (2.45 mM) were mixed under dark conditions for 16 h to generate ABTS radical cations. Then, 50 μL of gelatin peptide solution (6 mg/mL) or deionized water (as blank) was added to 150 μL of diluted ABTS solution (absorbance of 0.70 ± 0.02 at 734 nm) in a 96-well microplate. Subsequently, the solution was left to incubate for 6 min under dark conditions, and the absorbance was measured at a wavelength of 734 nm. The ABTS scavenging activity of gelatin peptides with different MWs was calculated as follows:

$$\text{ABTS radical scavening activity } (\%) = (1 - \frac{Asample}{Ablank}) \times 100$$

3.7. Peptide Sequence Identification Using Liquid Chromatography (LC)–Tandem Mass Spectrometry (MS/MS)

A Zorbax 300SB-C18 peptide trap (Agilent Technologies, Wilmington, DE, USA) was used to separate the gelatin peptides, and while a RP-C18 capillary column was used as the analytical column (0.15 mm × 150 mm; Column Technology Inc., Mississippi, USA). Separation was performed using a mobile phase consisting of 0.1% (v/v) formic acid aqueous solution (A) and 0.1% (v/v) formic acid acetonitrile solution (B), according to the method of Yu et al. [38]. The elution conditions were set as follows: 0–50 min, 4% B → 50% B; 50–54 min, 50% B → 100% B; and 54–60 min, 100% B. After the separation of gelatin peptides via capillary high-performance LC, they were analyzed via MS using a Q Exactive mass spectrometer (Thermo Fisher, Mass, USA). The analysis duration was set to 60 min, and detection was performed in the positive-ion mode. Ten fragment maps (MS2 scan) were collected after each full scan. Finally, the raw mass spectrum file was uploaded to MaxQuant 1.5.5.1 to obtain peptide sequences based on the database UniProt_Equus_asinus_33662_20221212. The peptide tolerance was set to 20 ppm, MS/MS tolerance was set to 0.1 Da, and false discovery rate of peptides and proteins was set to ≤0.01.

3.8. Peptide Synthesis

The antioxidant peptides with known sequences were synthesized by Hubei Qiangyao Biotech Co., Ltd. (Hubei, China), using Fmoc-protected amino acid synthesis.

3.9. Antioxidant Activity Tests of Synthetic Peptides

The antioxidant activity of synthetic peptides was tested using DPPH and ABTS radical scavenging assays, as described in Section 2.6. The concentration of synthetic peptides was determined to be 2 and 1 mg/mL according to DPPH and ABTS assays, respectively.

3.10. Molecular Docking of Gelatin Peptides with Keap1

The Keap1 protein with a three-dimensional crystal structure (obtained by ab initio modeling with AlphaFold2 [39]) was used as the initial receptor for molecular docking, and the PGPAP polypeptide was a straight chain structure directly constructed via Pymol [40]. The random coil part (1–55 amino acid residues) with very low modeling quality was deleted in the Keap1 sequence. This structure section is far from the overall active structure of the protein and has no impact on the biological function of the whole protein. The polypeptide and Keap1 were connected globally, and the sites with the strongest binding effect were selected for analysis. The binding site of the polypeptide molecule in a protein is primarily located in the cavity above the center of the barrel structure, which is composed of a large number of β-sheets at the C-terminus of the protein. Therefore, the active docking site of the polypeptide and protein was selected in this pocket. Molecular docking was implemented using AutoDock 4.2.6, and the center coordinates of the docking box were set as follows: x = 14.219, y = 1.850, and z = −21.613. The number of cells in each XYZ direction was set to 60 × 60 × 60, and the frequency of docking was set to 100 times. The docking was carried out by Lamarckian genetic algorithm in Autodock. The whole docking process adopted a semi-flexible method; that is, the receptor was regarded as rigid and the ligand as flexible. An independent docking run was performed during the molecular docking of gelatin peptides with Keap1. In this docking run, the protein and the small molecule were docked 50 times. An Amber14 force field was adopted for energy optimization [30]. It was carried out in two steps: firstly, the structure was optimized by the steepest descent method with 1000 steps, and then further optimized by the conjugate gradient method with 500 steps.

3.11. Molecular Dynamic (MD) Simulations of the PGPAP-Keap1 Complex

Gromacs 2018.4 [30] was used to run MD simulations of the PGPAP-Keap1 complex using the Amber14SB force field [41] and TIP3P water model [30]. During MD simulation, all bonds involving hydrogen atoms were constrained using the LINCS algorithm, and the integral step was 2 fsc. The electrostatic interaction was calculated using the particle-mesh Ewald method. The nonbond interaction cutoff value was set to 10 Å and was updated every 10 steps. The simulated temperature was controlled at 298.15 K using the V-rescale temperature coupling method, and the pressure was controlled at 1 bar using the Parrinello–Rahman method.

3.12. Statistical Analysis

One-way analysis of variance was used to perform triplicate analyses using SPSS 13.0 software (SPSS Inc., Chicago, IL, USA). All data were presented as the mean ± standard. A p-value of < 0.05 was considered to indicate statistical significance.

4. Conclusions

This study investigated the structural characteristics and antioxidant mechanism of donkey-hide gelatin peptides. The DPPH and ABTS free radical scavenging activity assays indicated that gelatin peptides with a MW of 1–3 kDa had the strongest antioxidant activity. Amino acid analysis, FTIR spectroscopy, and CD revealed that gelatin peptides with different MWs differed in terms of amino acid composition, peak strength of functional groups, and content of secondary structures, which might result in different antioxidant activities. In addition, molecular docking and MD simulation indicated that PGPAP could bind to the Keap1 protein at the following key residues: Arg415, Gly462, Phe478, and Tyr572. The PGPAP–Keap1 complex exhibited stable and compact conformations, which appear a plausible basis to exert its biological function. Overall, the antioxidant activity of gelatin peptides is closely related to their structural properties, and PGPAP may serve as an antioxidant peptide targeting in vivo metabolism.

Author Contributions: Conceptualization, R.L. and C.F.; methodology, R.L.; software, L.X.; validation, L.X.; formal analysis, L.X.; investigation, L.X.; resources, X.G.; data curation, L.C.; writing—original draft preparation, R.L.; writing—review and editing, C.F.; visualization, L.C.; supervision, X.G.; project administration, X.G.; funding acquisition, R.L. and X.G. All authors have read and agreed to the published version of the manuscript.

Funding: This research was funded by the Natural Science Foundation of Shandong Province (ZR2020QC220), Science and Technology Smes Innovation Ability Improvement Project of Shandong Province (2023TSGC0386; 2023TSGC0362), Key R&D Program of Shandong Province, China (2022TZXD0033), Higher Educational Youth Innovation Science and Technology Program of Shandong Province, China (2019KJF028) and Key Research and Development Plan Project of Liaocheng (2021NY05) are gratefully acknowledged.

Data Availability Statement: Data are contained within the article.

Conflicts of Interest: The data presented in this study are available on request from the corresponding author. The data are not publicly available due to privacy restrictions.

Abbreviations

Molecular weights, MW; 2, 2-Diphenyl-1-picrylhydrazyl, DPPH; 2, 2-azino-bis (3-ethylbenzothiazoline-6-sulfonic acid) diammonium salt, ABTS; Fourier transform infrared, FTIR; Circular dichroism, CD; Liquid chromatography−tandem mass spectrometry, LC−MS/MS; Molecular dynamic, MD; Potassium bromide, KBr.

References

1. Liang, R.; Cheng, S.; Dong, Y.; Ju, H. Intracellular antioxidant activity and apoptosis inhibition capacity of PEF-treated KDHCH in HepG2 cells. *Food Res. Int.* **2019**, *121*, 336–347. [CrossRef]
2. Xiang, N.; Lyu, Y.; Zhu, X.; Bhunia, A.K.; Narsimhan, G. Effect of physicochemical properties of peptides from soy protein on their antimicrobial activity. *Peptides* **2017**, *94*, 10–18. [CrossRef]
3. Tavakolipour, H. Extraction and evaluation of gelatin from silver carp waste. *World J. Fish Mar. Sci.* **2011**, *3*, 10–15.
4. Muyonga, J.H.; Cole, C.G.B.; Duodu, K.G. Fourier transform infrared (FTIR) spectroscopic study of acid soluble collagen and gelatin from skins and bones of young and adult Nile perch (*Lates niloticus*). *Food Chem.* **2004**, *86*, 325–332. [CrossRef]
5. Chi, C.F.; Wang, B.; Hu, F.Y.; Wang, Y.M.; Zhang, B.; Deng, S.G.; Wu, C.W. Purification and identification of three novel antioxidant peptides from protein hydrolysate of bluefin leatherjacket (*Navodon septentrionalis*) skin. *Food Res. Int.* **2015**, *73*, 124–129. [CrossRef]
6. Tkaczewska, J.; Bukowski, M.; Mak, P. Identification of antioxidant peptides in enzymatic hydrolysates of carp (*Cyprinus carpio*) skin gelatin. *Molecules* **2019**, *24*, 97. [CrossRef]
7. Ma, Y.; Wu, Y.; Li, L. Relationship between primary structure or spatial conformation and functional activity of antioxidant peptides from *Pinctada fucata*. *Food Chem.* **2018**, *264*, 108–117. [CrossRef]
8. Zhang, Y.; Zhang, H.; Wang, L.; Guo, X.; Qi, X.; Qian, H. Influence of the degree of hydrolysis (DH) on antioxidant properties and radical-scavenging activities of peanut peptides prepared from fermented peanut meal. *Eur. Food Res. Technol.* **2011**, *232*, 941–950. [CrossRef]
9. Liu, J.; Wang, X.; Zhao, Z. Effect of whey protein hydrolysates with different molecular weight on fatigue induced by swimming exercise in mice. *J. Sci. Food Agric.* **2014**, *94*, 126–130. [CrossRef] [PubMed]
10. Sun, Q.; Shen, H.; Luo, Y. Antioxidant activity of hydrolysates and peptide fractions derived from porcine hemoglobin. *J. Food Sci. Technol.* **2011**, *48*, 53–60. [CrossRef] [PubMed]
11. Bougatef, A.; Hajji, M.; Balti, R.; Lassoued, I.; Triki-Ellouz, Y.; Nasri, M. Antioxidant and free radical-scavenging activities of smooth hound (*Mustelus mustelus*) muscle protein hydrolysates obtained by gastrointestinal proteases. *Food Chem.* **2009**, *114*, 1198–1205. [CrossRef]
12. Laakso, S. Inhibition of lipid peroxidation by casein Evidence of molecular encapsulation of 1, 4-pentadiene fatty acids. *Biochim. Et Biophys. Acta (BBA)-Lipids Lipid Metab.* **1984**, *792*, 11–15. [CrossRef]
13. Leung, C.H.; Zhang, J.T.; Yang, G.J.; Liu, H.; Han, Q.B.; Ma, D.L. Emerging screening approaches in the development of Nrf2-Keap1 protein-protein interaction inhibitors. *Int. J. Mol. Sci.* **2019**, *20*, 4445. [CrossRef]
14. Li, L.; Liu, J.; Nie, S.; Ding, L.; Wang, L.; Liu, J.; Liu, W.; Zhang, T. Direct inhibition of Keap1-Nrf2 interaction by egg-derived peptides DKK and DDW revealed by MD and fluorescence polarization. *RSC Adv.* **2017**, *7*, 34963–34971. [CrossRef]

15. Agrawal, H.; Joshi, R.; Gupta, M. Purification, identification and characterization of two novel antioxidant peptides from finger millet (Eleusine coracana) protein hydrolysate. *Food Res. Int.* **2019**, *120*, 697–707. [CrossRef]
16. Tonolo, F.; Fiorese, F.; Moretto, L.; Folda, A.; Scalcon, V.; Grinzato, A.; Ferro, S.; Arrigoni, G.; Bindoli, A.; Feller, E.; et al. Identification of new peptides from fermented milk showing antioxidant properties: Mechanism of action. *Antioxidants* **2020**, *9*, 117. [CrossRef]
17. Tonolo, F.; Moretto, L.; Grinzato, A.; Fiorese, F.; Folda, A.; Scalcon, V.; Ferro, S.; Arrigoni, G.; Bellamio, M.; Feller, E.; et al. Fomented soy-derived bioactive peptides selected by a molecular docking approach show antioxidant properties involving the Keap1/Nrf2 pathway. *Antioxidants* **2020**, *9*, 1306. [CrossRef]
18. Ngoh, Y.; Gan, C. Enzyme-assisted extraction and identification of antioxidative and α-amylase inhibitory peptides from Pinto beans (*Phaseolus vulgaris* cv. Pinto). *Food Chem.* **2016**, *190*, 331–337. [CrossRef] [PubMed]
19. Dávalos, A.; Miguel, M.; Bartolome, B.; López-Fandiño, R. Antioxidant activity of peptides derived from egg white proteins by enzymatic hydrolysis. *J. Food Prot.* **2004**, *67*, 1939–1944. [CrossRef] [PubMed]
20. Sarmadi, B.H.; Ismail, A. Antioxidative peptides from food proteins: A review. *Peptides* **2010**, *31*, 1949–1956. [CrossRef] [PubMed]
21. Suetsuna, K.; Ukeda, H.; Ochi, H. Isolation and characterization of free radical scavenging activities peptides derived from casein. *J. Nutr. Biochem.* **2000**, *11*, 128–131. [CrossRef]
22. Chi, C.F.; Hu, F.Y.; Wang, B.; Ren, X.J.; Deng, S.G.; Wu, C.W. Purification and characterization of three antioxidant peptides from protein hydrolyzate of croceine croaker (*Pseudosciaena crocea*) muscle. *Food Chem.* **2015**, *168*, 662–667. [CrossRef]
23. Saito, K.; Jin, D.H.; Ogawa, T.; Muramoto, K.; Hatakeyama, E.; Yasuhara, T.; Nokihara, K. Antioxidative properties of tripeptide libraries prepared by the combinatorial chemistry. *J. Agric. Food Chem.* **2003**, *51*, 3668–3674. [CrossRef]
24. Tamburro, A.M.; Bochicchio, B.; Pepe, A. Dissection of human tropoelastin: Exon-by-exon chemical synthesis and related conformational studies. *Biochemistry* **2003**, *42*, 13347–13362. [CrossRef]
25. Yang, R.; Wang, J.; Lin, S.; Ye, H.; Chen, F. In vitro antioxidant activities of the novel pentapeptides Ser-His-Glu-Cys-Asn and Leu-Pro-Phe-Ala-Met and the relationship between activity and peptide secondary structure. *J. Sci. Food Agric.* **2017**, *97*, 1945–1952. [CrossRef]
26. Liang, R.; Zhang, Z.; Lin, S. Effects of pulsed electric field on intracellular antioxidant activity and antioxidant enzyme regulating capacities of pine nut (*Pinus koraiensis*) peptide QDHCH in HepG2 cells. *Food Chem.* **2017**, *237*, 793–802. [CrossRef]
27. Liang, R.; Cheng, S.; Lin, S.; Dong, Y.; Ju, H. Validation of Steric Configuration Changes Induced by a Pulsed Electric Field Treatment as the Mechanism for the Antioxidant Activity Enhancement of a Peptide. *Food Bioprocess Technol.* **2021**, *14*, 1751–1757. [CrossRef]
28. Chen, H.M.; Muramoto, K.; Yamauchi, F.; Nokihara, K. Antioxidant activity of designed peptides based on the antioxidative peptide isolated from digests of a soybean protein. *J. Agric. Food Chem.* **1996**, *9*, 44. [CrossRef]
29. Xu, L.; Li, X.; Huang, Y.; Wu, X.; Hou, R.; Wang, H.; Wang, N.; Zhang, X. Protection of small molecule corn peptide Leu-Asp-Tyr-Glu from mitochondria against oxidative damage. *Chem. J. Chin. Univ.* **2004**, *25*, 1073–1075. Available online: http://www.cjcu.jlu.edu.cn/EN/Y2004/V25/I6/1073 (accessed on 6 June 2023).
30. Maier, J.A.; Martinez, C.; Kasavajhala, K.; Wickstrom, L.; Hauser, K.E.; Simmerling, C. ff14SB: Improving the accuracy of protein side chain and backbone parameters from ff99SB. *J. Chem. Theory Comput.* **2015**, *11*, 3696–3713. [CrossRef] [PubMed]
31. Zhang, D.D. The Nrf2-Keap1-ARE signaling pathway: The regulation and dual function of Nrf2 in cancer. *Antioxid. Redox Signal.* **2010**, *13*, 1623. [CrossRef] [PubMed]
32. Awuh, J.A.; Haug, M.; Mildenberger, J.; Marstad, A.; Do, C.P.N.; Louet, C.; Stenvik, J.; Steigedal, M.; Damås, J.K.; Halaas, Ø. Keap1 regulates inflammatory signaling in Mycobacterium avium-infected human macrophages. *Proc. Natl. Acad. Sci. USA* **2015**, *112*, 4272–4280. [CrossRef] [PubMed]
33. Greenwald, M.B.Y.; Frušić-Zlotkin, M.; Soroka, Y.; Sasson, S.B.; Bianco-Peled, H.; Bitton, R.; Kohen, R. Nitroxide delivery system for Nrf2 activation and skin protection. *Eur. J. Pharm. Biopharm.* **2015**, *94*, 123–134. [CrossRef] [PubMed]
34. Hayes, J.D.; Mcmahon, M.; Chowdhry, S.; Dinkova-Kostova, A.T. Cancer chemoprevention mechanisms mediated through the Keap1-Nrf2 pathway. *Antioxid. Redox Signal.* **2010**, *13*, 1713–1748. [CrossRef] [PubMed]
35. Yu, Z.; Wang, Y.; Shuian, D.; Liu, J.; Zhao, W. Identification and molecular mechanism of novel immunomodulatory peptides from gelatin hydrolysates: Molecular docking, dynamic simulation, and cell experiments. *J. Agric. Food Chem.* **2023**, *71*, 2924–2934. [CrossRef]
36. Provencher, S.W.; Gloeckner, J. Estimation of globular protein secondary structure from circular dichroism. *Biochemistry* **1981**, *20*, 33–37. [CrossRef] [PubMed]
37. Kozics, K.; Klusová, V.; Srancíková, A.; Mucaji, P.; Slamenová, D.; Hunáková, L.; Kusznierewicz, B.; Horváthová, E. Effects of Salvia officinalis and Thymus vulgaris on oxidant-induced DNA damage and antioxidant status in HepG2 cells. *Food Chem.* **2013**, *141*, 2198–2206. [CrossRef]
38. Yu, Z.; Kan, R.; Ji, H.; Wu, S.; Zhao, W.; Shuian, D.; Liu, J.; Li, J. Identification of tuna protein-derived peptides as potent SARSCoV-2 inhibitors via molecular docking and molecular dynamic simulation. *Food Chem.* **2021**, *342*, 128366. [CrossRef]
39. Jumper, J.; Evans, R.; Pritzel, A.; Green, T.; Figurnov, M.; Ronneberger, O.; Tunyasuvunakool, K.; Bates, R.; Žídek, A.; Potapenko, A.; et al. Highly accurate protein structure prediction with AlphaFold. *Nature* **2021**, *596*, 583–589. [CrossRef]

40. Lill, M.A.; Danielson, M.L. Computer-aided drug design platform using PyMOL. *J. Comput.-Aided Mol. Des.* **2011**, *25*, 13–19. [CrossRef]
41. Spoel, D.V.D.; Lindahl, E.; Hess, B.; Groenhof, G.; Mark, A.E.; Berendsen, H.J.C. GROMACS: Fast, flexible, and free. *J. Comput. Chem.* **2005**, *26*, 1701–1718. [CrossRef] [PubMed]

molecules

MDPI

Article

Effects of Ultra-High-Pressure Jet Processing on Casein Structure and Curdling Properties of Skimmed Bovine Milk

Fei Xu [1,2], Lu Xue [1,*], Yanfeng Ma [2], Tianjiao Niu [2], Pei Zhao [1], Zijian Wu [1] and Yanfa Wang [2,*]

1 Tianjin Key Laboratory of Food Biotechnology, College of Biotechnology and Food Science, Tianjin University of Commerce, Tianjin 300134, China
2 Mengniu Hi-Tech Dairy (Beijing) Co., Ltd., Beijing 101107, China
* Correspondence: xuelu@tjcu.edu.cn (L.X.); wangyanfa@mengniu.cn (Y.W.)

Abstract: Ultra-high-pressure jet processing (UHPJ) is a new non-thermal processing technique that can be employed for the homogenization and the sterilization of dairy products. However, the effects on dairy products are unknown when using UHPJ for homogenization and sterilization. Thus, this study aimed to investigate the effects of UHPJ on the sensory and curdling properties of skimmed milk and the casein structure in skimmed milk. Skimmed bovine milk was treated with UHPJ using different pressures (100, 150, 200, 250, 300 MPa) and casein was extracted by isoelectric precipitation. Subsequently, the average particle size, Zeta potential, contents of free sulfhydryl and disulfide bonds, secondary structure, and surface micromorphology were all used as evaluation indicators to explore the effects of UHPJ on the structure of casein. The results showed that with an increase of pressure, the free sulfhydryl group content changed irregularly, while the disulfide bond content increased from 1.085 to 3.0944 μmol/g. The content of α-helix and random coil in the casein decreased, while the β-sheet content increased at 100, 150, 200 MPa pressure. However, treatment with higher pressures of 250 and 300 MPa had the opposite effect. The average particle size of the casein micelles first decreased to 167.47 nm and then increased up to 174.63 nm; the absolute value of Zeta potential decreased from 28.33 to 23.77 mV. Scanning electron microscopy analysis revealed that the casein micelles had fractured into flat, loose, porous structures under pressure instead of into large clusters. After being ultra-high-pressure jet-processed, the sensory properties of skimmed milk and its fermented curd were analyzed concurrently. The results demonstrated that UHPJ could alter the viscosity and color of skimmed milk, shortening curdling time from 4.5 h to 2.67 h, and that the texture of the curd fermented with this skimmed milk could be improved to varying degrees by changing the structure of casein. Thus, UHPJ has a promising application in the manufacture of fermented milk due to its ability to enhance the curdling efficiency of skimmed milk and improve the texture of fermented milk.

Keywords: ultra-high-pressure jet processing; casein; protein structure; curdling properties

Citation: Xu, F.; Xue, L.; Ma, Y.; Niu, T.; Zhao, P.; Wu, Z.; Wang, Y. Effects of Ultra-High-Pressure Jet Processing on Casein Structure and Curdling Properties of Skimmed Bovine Milk. *Molecules* **2023**, *28*, 2396. https://doi.org/10.3390/molecules28052396

Academic Editors: Shudong He, Wei Xu and Muhammad H. Aludatt

Received: 29 December 2022
Revised: 26 February 2023
Accepted: 27 February 2023
Published: 6 March 2023

1. Introduction

Ultra-high-pressure jet (UHPJ) processing is a new non-thermal processing technology. It divides liquids into two or more streams for Y-type, Z-type, or vertical impact [1]. The impact process combines multiple mechanical forces, such as high-velocity impact, high-frequency vibration, instantaneous pressure drop, intense shear, cavitation, and ultra-high pressure, and can generate a dynamic pressure of up to 200 MPa in a short period. Consequently, UHPJ processing can facilitate the homogenization of liquid materials, physical modification, and auxiliary sterilization.

Studies have been conducted to investigate the effects of UHPJ processing on the structural and functional characteristics of proteins derived from food, such as ovalbumin [2], isolated soybean protein [3], peanut protein isolate [4], and myofibrillar protein [5]. The results indicate that UHPJ processing conducted at a relatively low pressure can partially unfold and denature part of a protein structure, leading to a decrease in the particle size and a changing in its secondary structure. As a result, proteins transition from an ordered state to a disordered one, while their structure loosens and the surface hydrophobicity increases. With an increase of pressure or treatment time, a protein will refold and aggregate, resulting in an increase in the particle size and a decrease in the surface hydrophobicity. In addition, UHPJ processing can enhance protein solubility [6]. At present, the research is focused on the alteration of high-purity food-derived proteins using UHPJ processing, though a few studies have explored the changes of unpurified proteins in raw materials using UHPJ processing.

Bovine milk, which is rich in protein, fat, vitamins, and minerals, is an important source of nutrition for humans. Accounting for approximately 80% of its total protein content, casein is the most abundant protein in milk and is divided into four main types: αs_1-casein, αs_2-casein, β-casein, and κ-casein [7]. Given that UHPJ processing can be applied to dairy manufacture [8,9], it is necessary to investigate the changes of casein in milk after UHPJ processing, and the potential effects of such processing on subsequent dairy processing. In this study, skimmed milk was treated using UHPJ processing at 100–300 MPa. Subsequently, the effects of the pressure on the structure of casein in skimmed milk and the changes of the sensory properties of skimmed milk and skimmed milk curd were studied. This provided a theoretical reference for the application of UHPJ technology in dairy processing.

2. Results and Discussion

2.1. Effect of UHPJ Processing on Temperature of Skimmed Milk

As shown in Figure 1, the sampling temperature of skimmed milk rose rapidly with the increase in pressure. The injection temperature of the skimmed milk in this study was 10 °C. When the treatment pressure increased from 100 MPa to 300 MPa, the sampling temperature gradually increased from 39.5 ± 2.9 °C to 83.4 ± 3.2 °C. The reasons for this increase in temperature include ultra-high pressure, intense friction caused by shear, and the vaporization of milk when forced through a narrow reaction chamber microchannel [10]. This demonstrates that the UHPJ processing can increase the temperature of skimmed milk, which may be the consequence of high-velocity impact, high-frequency vibration, intense shear, or cavitation in addition to adiabatic heating generated by the machine [11].

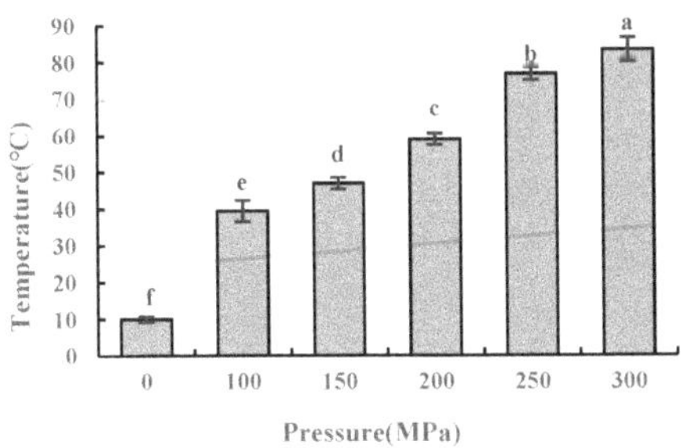

Figure 1. Sampling temperature of skimmed milk under different pressure conditions. Different superscript letters within a column indicate significant differences between means ($p < 0.05$).

2.2. Effect of UHPJ Processing on the Composition of Skimmed Milk

The composition of skimmed milk was determined by the Milko Scan FT-120. The data in Table 1 shows that the UHPJ processing had no significant effect on the composition of the skimmed milk, thus indicating that the UHPJ processing will not cause the decomposition of nutrients.

Table 1. The composition of samples.

	Protein (%)	Fat (%)	Lactose (%)	SNF (%)	TS (%)
Raw milk	3.37 ± 0.01	4.04 ± 0.02	4.73 ± 0.01	9.18 ± 0.02	12.98 ± 0.02
Skimmed milk	3.01 ± 0.00 [a]	0.11 ± 0.03 [a]	4.91 ± 0.00 [a]	9.76 ± 0.01 [a]	9.82 ± 0.02 [a]
100 MPa	3.01 ± 0.02 [a]	0.11 ± 0.02 [a]	4.91 ± 0.03 [a]	9.76 ± 0.02 [a]	9.82 ± 0.02 [a]
150 MPa	3.01 ± 0.03 [a]	0.11 ± 0.01 [a]	4.91 ± 0.02 [a]	9.76 ± 0.01 [a]	9.82 ± 0.02 [a]
200 MPa	3.01 ± 0.02 [a]	0.11 ± 0.02 [a]	4.91 ± 0.01 [a]	9.76 ± 0.02 [a]	9.82 ± 0.02 [a]
250 MPa	3.01 ± 0.02 [a]	0.11 ± 0.02 [a]	4.91 ± 0.02 [a]	9.76 ± 0.02 [a]	9.82 ± 0.02 [a]
300 MPa	3.01 ± 0.02 [a]	0.11 ± 0.02 [a]	4.91 ± 0.02 [a]	9.76 ± 0.03 [a]	9.82 ± 0.02 [a]

Results are presented as mean ± SD for triplicate samples. SNF: non-fat milk solid. TS: total solids. Different superscript letters within a column indicate significant differences between means ($p < 0.05$).

2.3. *Effect of UHPJ Processing on Content of Free Sulfhydryl and Disulfide Bonds in Casein*

Free sulfhydryl and disulfide bonds are important chemical bonds that stabilize the conformation of protein molecules and are critical in determining the functional properties of proteins [12]. Treatments such as high pressure and heating can cause changes in the content of the free sulfhydryl and disulfide bonds in proteins. As shown in Figures 2 and 3), the content of the free sulfhydryl in casein increased initially but then decreased after the high-pressure jet treatment. This content was significantly higher than that of the control group except for the sample at 200 MPa ($p < 0.05$). The content of the disulfide bond decreased at first but then increased and was significantly higher than that of the control group ($p < 0.05$) except for the 100 MPa sample; this is consistent with results from Wang et al. [13]. Under 100 MPa pressure, the content of the free sulfhydryl group in casein was 0.6887 μmol/g, which was significantly higher than that of the control group ($p < 0.05$). This may be attributed the strong shear and impact of the high-pressure jet, which destroyed the molecular structure of casein, exposing the sulfhydryl groups buried in αs_2-casein [14] and κ-casein [15] molecules. The production rate of the free sulfhydryl group was higher than that of the oxidation, thus resulting in an increase in the content of the free sulfhydryl group ($p < 0.05$) and a decrease in the content of the disulfide bond ($p < 0.05$). The oxidation of the sulfhydryl groups was dominant at 150–300 MPa. The surface of the newly formed casein particles was thermodynamically unstable and some of the sulfhydryl groups were folded and embedded into the casein molecules. Therefore, compared with a treatment pressure of 100 MPa, the free sulfhydryl group content of casein decreased significantly ($p < 0.05$) while the disulfide bond content increased significantly ($p < 0.05$).

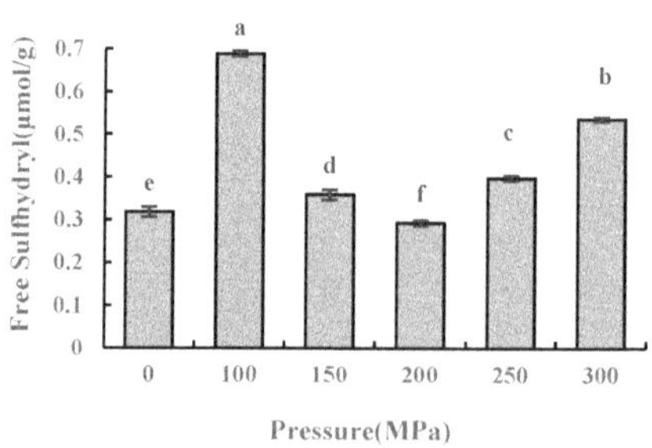

Figure 2. Free sulfhydryl content of casein under different pressure conditions. Different superscript letters within a column indicate significant differences between means ($p < 0.05$).

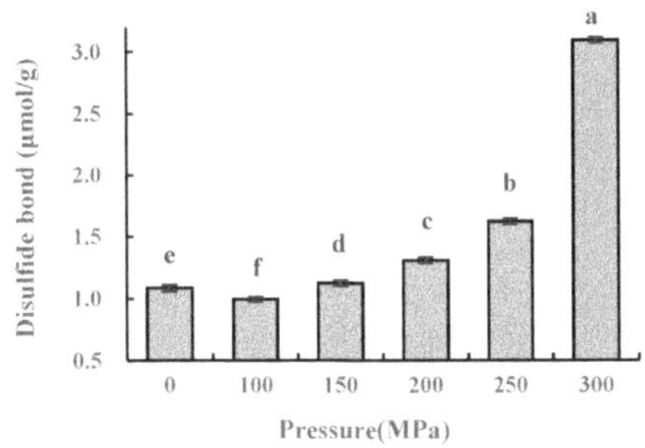

Figure 3. Disulfide bond content of casein under different pressure conditions. Different superscript letters within a column indicate significant differences between means ($p < 0.05$).

The content of the disulfide bond in casein increased more under 200–300 MPa pressure, which may be attributed to the denaturation of β-lactoglobulin owing to the higher pressure and temperature, and then formed an intermolecular disulfide bond with κ-casein [16]. The content of the disulfide bond reached its maximum value at 300 MPa, and simultaneously, the sensory evaluation experiment results showed that the cooking flavor of skimmed milk was also at its strongest (data not shown), indicating that an excessively high pressure was not conducive to preserving the milk's unique flavor.

2.4. Effect of UHPJ Processing on Secondary Structure of Casein

The direction of the hydrogen bond in an α-helix is consistent with that of the helix axis, consequently it has strong rigidity, while the direction of hydrogen bond in a β-sheet is perpendicular to the direction of folding, consequently it has better flexibility. The stability of these secondary structures is different when subjected to the UHPJ processing and thermal treatment. The effect of the UHPJ processing on the content of the casein secondary structure is shown in Table 2. After pressure treatments of 100, 150, and 200 MPa, the content of the α-helix decreased significantly ($p < 0.05$), which may be attributed to the fact that the hydrogen bonds which maintain the structural stability of casein are partially broken under the action of a high-pressure jet, leading to the extension of the casein. Such a significant increase in the content of the α-helix ($p < 0.05$) under 250 and 300 MPa pressure may be due to the synergistic effect of high temperature and pressure, which enhances the interaction between casein molecules, changing the orientation of the hydrogen bonds in casein to generate a greater amount of them [17]. There was no significant difference in the content of the β-turn ($p > 0.05$) under different pressure conditions. The content of the random coil first decreased and then increased, while the content of the β-sheet changed in the opposite direction. In summary, the content of the α-helix and the random coil of casein decreased, yet the content of the β-sheet increased, when treated with the UHPJ at lower pressures (100, 150, 200 MPa), while the effect of UHPJ treatment with a higher pressure (250, 300 MPa) had the opposite effect.

Table 2. Content of secondary structural components of casein under different pressure conditions.

Pressure/MPa	Content of Secondary Structure Components of Casein (%)			
	α-Helic	β-Sheet	β-Turn	Random Coils
0	12.22 ± 0.23 [c]	26.92 ± 0.35 [b]	25.07 ± 0.04 [a]	36.14 ± 0.22 [c]
100	11.14 ± 0.18 [f]	27.35 ± 0.11 [b]	24.86 ± 0.08 [a]	34.75 ± 0.17 [e]
150	11.73 ± 0.32 [d]	28.41 ± 0.25 [a]	25.19 ± 0.12 [a]	35.09 ± 0.13 [de]
200	11.48 ± 0.09 [e]	28.95 ± 0.05 [a]	24.65 ± 0.22 [a]	35.52 ± 0.19 [d]
250	14.39 ± 0.15 [a]	24.63 ± 0.18 [c]	24.53 ± 0.14 [a]	37.07 ± 0.16 [b]
300	13.21 ± 0.21 [b]	23.18 ± 0.16 [d]	24.71 ± 0.13 [a]	39.03 ± 0.21 [a]

Results are presented as mean ± SD for triplicate samples; different superscript letters within a column indicate significant differences between means ($p < 0.05$).

2.5. Effect of UHPJ Processing on Average Particle Size of Casein Micelle

The average particle size of the casein micelles under different pressures is shown in Figure 4. The average particle size decreased first but then increased with an increase in the treatment pressure, which is consistent with the research results from Mohan et al. [18]. At the relatively low pressures of 100 MPa and 150 MPa, the decrease in average casein particle size was due to external forces such as impact force, strong shear, and the void effect acting on the casein in the reaction chamber of the UHPJ homogenizer. The casein particles were crushed and the micelle structure was destroyed such that the average particle size was reduced. The average particle size of the casein micelles increased under pressures of 200–300 MPa, with no significant difference in average particle size between the sample at 300 MPa and the control. The reasons for this phenomenon are as follows: (1) The dissolution of κ-casein on the surface of the casein micelles under high pressure reduced the spatial repulsion between the micelles and the negative charge on the surface, thus promoting aggregation between the micelles [19]. (2) In the process of using the UHPJ, an increase of pressure was accompanied by an increase of temperature. β-lactoglobulin will denature when the pressure reaches 200 MPa or the temperature is 75–80 °C. As shown in Figure 1, the sample temperature of skimmed milk under 250 MPa pressure treatment reached 76.9 °C. The modified β-lactoglobulin interacted with the surface of the casein micelles, binding to increase the average particle size of the micelles [20].

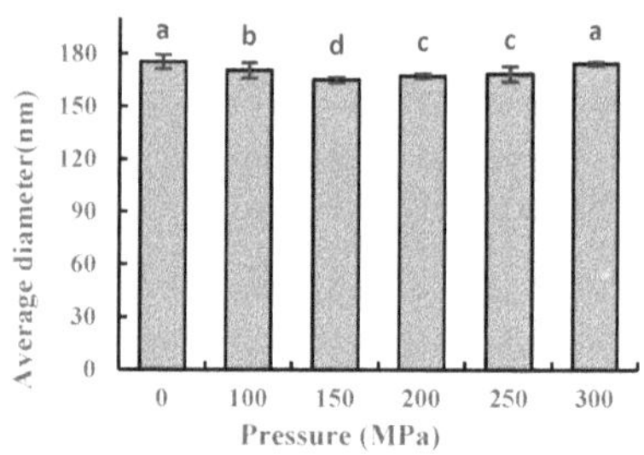

Figure 4. Average particle size of casein under different pressure conditions. Different superscript letters within a column indicate significant differences between means ($p < 0.05$).

2.6. Effect of UHPJ Processing on Zeta Potential of Casein Micelles

Zeta potential is the potential of the shear surface in a charged double layer of charged particles in solution [21], and it is one of the indicators used to evaluate the stability of colloids. The greater the absolute value of the Zeta potential, the greater the electrostatic repulsion force on the molecular surface and the more stable the casein colloid in the milk system. The smaller the absolute value of the Zeta potential, the more unstable the casein micelle. The absolute value of the Zeta potential of the samples in this study decreased after the pressure treatment compared with the control group, indicating that the stability of casein micelles also decreased (Figure 5). Janahar et al. [22] found that pressure-only treatments up to 400 MPa do not reduce the apparent particle size of whole milk's viscosity, while ultra-shear technology (UST) or high-pressure homogenization facilitate both the particle size and the Zeta potential reduction. This suggests that shear is the dominant effect for the change of the particle size in ultra-high pressure processing.

Sandra [23] reported that the mechanism by which ultra-high-pressure homogenization (UHPH) modifies the structural properties of casein micelles in reconstituted skimmed milk powder is likely to be due to hydrophobic interaction or a shearing effect, while the application of heat treatment only has a minimal effect on the average casein micelle size. UHPH does not disrupt the casein micelles completely but rather dissociates parts of their surfaces. As a result, the protein complexes formed after UHPH are different from those formed after heating.

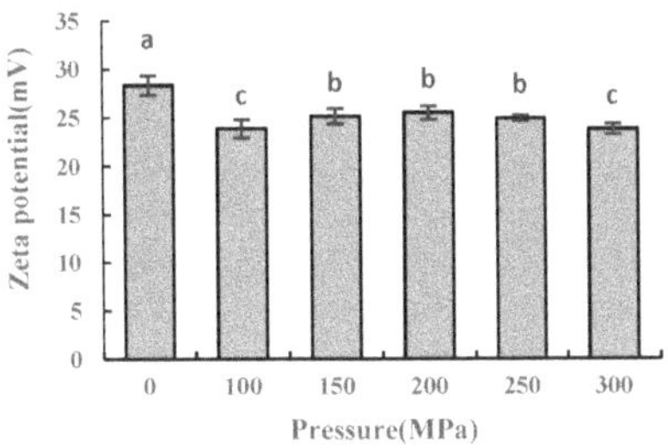

Figure 5. Zeta potential of casein under different pressure conditions. Different superscript letters within a column indicate significant differences between means ($p < 0.05$).

2.7. Effect of UHPJ Processing on the Microstructure of Casein

The microstructure of casein treated by UHPJ processing was observed by scanning electron microscope and the results are shown in Figure 6. Casein without the UHPJ processing showed an irregular and large cluster structure (Figure 6A). After being treated at 100 MPa (Figure 6B) and 150 MPa (Figure 6C) respectively, large clusters of casein were dispersed into small fluffy and porous clumps, which is consistent with the results of the particle size reduction. This shows that UHPJ processing can destroy protein structures and produce fragments of differing size and shape [24]. When the pressure exceeded 200 MPa, the casein micelle showed completely different structural characteristics, such as the formation of small casein clumps into aggregates. Casein aggregates formed at 250 MPa (Figure 6D) were larger than those formed at 200 MPa (Figure 6E), which is consistent with the results of the particle size increase. When the pressure reached 300 MPa, dense casein aggregates formed and the micelle structure flattened (Figure 6F). The above results suggest that protein depolymerization can be promoted by UHPJ processing at lower pressures, whereas protein re-aggregation can be initiated at higher pressures. Correspondingly, the casein particle size decreased at first and then increased, and the minimum particle size was observed at 150 MPa.

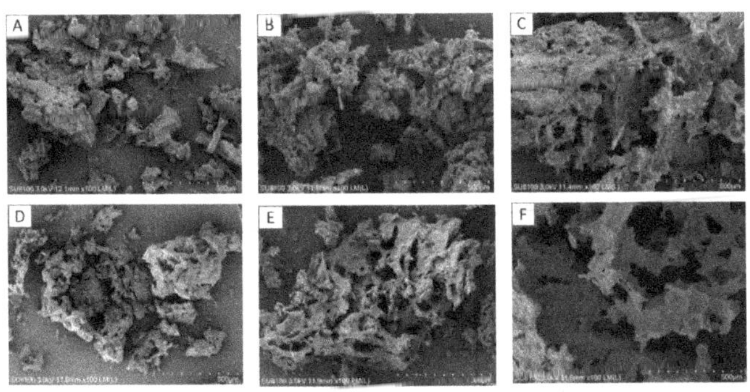

Figure 6. Scanning electron microscope diagram of casein under different pressure conditions. (**A**): 0 MPa (control); (**B**): 100 MPa; (**C**): 150 MPa; (**D**): 200 MPa; (**E**): 250 MPa; (**F**): 300 MPa; The magnification of scanning electron microscope was 100 times.

In addition, it has been demonstrated that colloidal calcium phosphate plays a critical role in maintaining the structural stability of the casein micelles [25]. High pressure can lead to the dissolution of colloidal calcium phosphate and a decrease in the particle size of the casein micelles.

2.8. Effect of UHPJ Processing on Apparent Viscosity of Skimmed Milk

The change in apparent viscosity of the skimmed milk after the UHPJ processing is shown in Figure 7. The variation trend of apparent viscosity with pressure is consistent with that of the average particle size (Figure 4). Janahar et al. [22] reported that there is no significant difference between the viscosity of 400 MPa ultra-shear technology-treated whole milk and that of an untreated sample. Their results are similar to the viscosity of the sample treated with 300 MPa UHPJ in our experiment, although skimmed milk was used in our experiment. This phenomenon can be attributed to the disintegration of the original casein micelles under high pressure, which resulted in the formation of larger protein aggregates with a higher hydrodynamic volume [26].

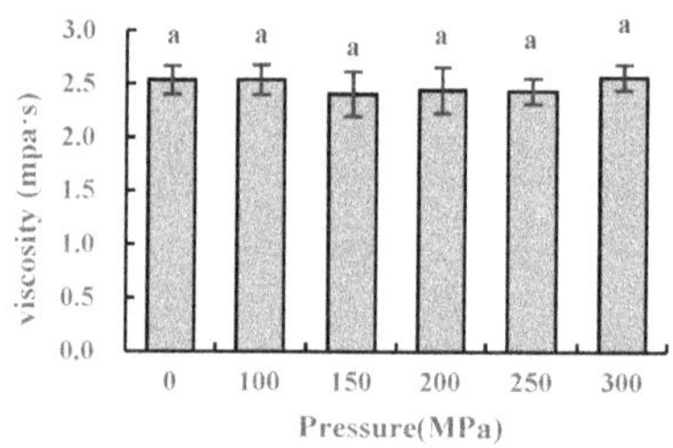

Figure 7. Apparent viscosity of skimmed milk under different pressure conditions. Different superscript letters within a column indicate significant differences between means ($p < 0.05$).

2.9. Effect of UHPJ Processing on Skimmed Milk Color

The color of the object is typically expressed using a three-dimensional color model of *L-a-b* or L^*-a^*-b^*, wherein L^* represents brightness; a^* represents the red-green system, with $a^* < 0$ indicating the sample is greener than the standard; b^* represents the yellow-blue system, with $b^* < 0$ indicating the sample is bluer than the standard. At present, the color of milk and dairy products is mainly analyzed according to the relationship between the changes of L^* and the changes of the milk protein structure, especially the particle size of the casein micelles. The color change of skimmed milk after the UHPJ processing is shown in Table 2. The smaller the value of L^*, the lower the brightness of the skimmed milk, where L^* is related to the state and the particle size of the casein. After being treated with UHPJ, the L^* value of the skimmed milk decreased initially (100, 150 MPa) and subsequently increased (200–300 MPa). Usually, the variation of L^* was consistent with that of the average particle size, which may be attributed to the destruction of the casein micelles and the formation of small fragments, thus increasing the transmittance of the milk [27,28]. In this study, there was no significant difference among all the samples ($p > 0.05$), which means the UHPJ processing had no significant effect on the brightness of skimmed milk.

The values of a^* and b^* for the samples both showed the trend of decreasing initially and then increasing, which is consistent with results from Pereda et al. [11]. This means that samples were much greener and bluer than the control under the pressures of 100–250 MPa, while samples were redder and yellower than the control at 300 MPa. However, there are few studies on the relationship between the change of a^* and b^* and the specific structural change of the components in skimmed milk.

The chromatic aberration value ΔE^* presents overall color change, with a higher value of ΔE^* indicating a greater overall color change between the sample and the control. According to the ΔE^* data in Table 3, the color of all the samples changed to some extent, among which skimmed milk at 100 MPa showed the greatest difference in color compared to the control group.

Table 3. Color change of skimmed milk under different pressures.

Pressure/MPa	L^*	a^*	b^*	ΔE^*
0	70.60 ± 0.29^{a}	-7.22 ± 0.07^{a}	-5.12 ± 0.07^{b}	/
100	70.61 ± 0.29^{a}	-7.79 ± 0.27^{b}	-6.57 ± 0.27^{d}	1.57 ± 0.31^{a}
150	70.80 ± 0.12^{a}	-7.74 ± 0.05^{b}	-5.74 ± 0.12^{c}	0.90 ± 0.00^{b}
200	70.50 ± 0.25^{a}	-8.01 ± 0.10^{b}	-5.55 ± 0.30^{bc}	0.91 ± 0.27^{b}
250	70.60 ± 0.33^{a}	-7.87 ± 0.06^{b}	-5.33 ± 0.06^{bc}	0.68 ± 0.02^{b}
300	70.59 ± 0.39^{a}	-6.89 ± 0.13^{a}	-4.16 ± 0.22^{a}	1.03 ± 0.20^{b}

Results are presented as mean ± SD for triplicate samples; different superscript letters within a column indicate significant differences between means ($p < 0.05$).

2.10. Effect of UHPJ Processing on Coagulation Properties of Skimmed Milk

The particle size (Figure 4), secondary structure (Table 2), and microstructure (Figure 6) of the casein changed after the UHPJ processing; therefore, the processing properties of raw milk [28–31], such as the coagulation properties, emulsibility, and foaming properties, changed accordingly. In order to investigate the effects of the UHPJ processing on the follow-up processing of dairy products, skimmed yogurt was prepared from the skimmed milk that had been subjected to the UHPJ processing. The indexes related to the quality of the yogurt, such as the clotting time, apparent viscosity, and the water holding capacity, were evaluated and the results are shown in Table 4.

Table 4. Coagulation properties of skimmed milk after different pressure treatments.

Pressure/MPa	Curding Time/h	Apparent Viscosity/mPa·s	Water Holding Capacity/%	Firmness/g	Cohesive/g	Viscosity/g.Sec
0	4.50 ± 0.08^{a}	70.72 ± 0.12^{f}	24.51 ± 1.1^{d}	6.21 ± 0.56^{d}	0.79 ± 0.53^{d}	1.19 ± 0.78^{e}
100	4.50 ± 0.22^{a}	131.93 ± 0.16^{d}	25.77 ± 0.95^{c}	6.36 ± 0.37^{c}	0.81 ± 0.42^{bc}	1.21 ± 0.43^{cd}
150	3.75 ± 0.15^{b}	130.81 ± 0.09^{de}	26.86 ± 1.5^{b}	6.35 ± 1.2^{c}	0.87 ± 0.86^{b}	1.39 ± 0.39^{c}
200	3.33 ± 0.05^{bc}	148.16 ± 0.13^{c}	26.80 ± 1.3^{b}	6.33 ± 0.81^{c}	0.85 ± 0.94^{b}	1.43 ± 1.2^{c}
250	2.67 ± 0.8^{d}	286.38 ± 0.21^{b}	35.31 ± 0.68^{a}	6.86 ± 0.85^{a}	1.07 ± 0.68^{a}	1.95 ± 0.94^{b}
300	2.67 ± 0.13^{d}	419.95 ± 0.25^{a}	35.85 ± 0.84^{a}	6.61 ± 0.62^{b}	1.10 ± 0.83^{a}	2.39 ± 1.7^{a}

Results are presented as mean ± SD for triplicate samples; different superscript letters within a column indicate significant differences between means ($p < 0.05$).

UHPJ processing above 150 MPa can shorten the curdling time of skimmed milk to varying degrees; for example, the clotting time of the samples treated with 250 and 300 MPa was 1.83 h shorter than that of the control group. In contrast to the change trend in the apparent viscosity of the unfermented skimmed milk treated by the UHPJ processing (Figure 7), the apparent viscosity and the viscosity of yoghurt prepared from this skimmed milk always increased with the increase of pressure. The water holding capacity (WHC) represents the ability of curd to retain water under the action of centrifugal, stirring, and other external forces. The WHC of skimmed yogurt increased continuously with an increase of pressure and reached its maximum when the pressure was 250 MPa, as reported by Serra et al. [28] and Ciron et al. [32]. The strength of the sample's intermolecular binding was reflected by its cohesiveness, which increased as the pressure increased, suggesting that the UHPJ processing enhanced the interaction of the proteins in the skimmed yogurt. Firmness also kept rising with an increase in the pressure. When the pressure reached 250 MPa, however, it attained its extreme value. Even though firmness slightly declined when the pressure reached 300 MPa, it was still significantly higher than that of the control group.

The changes in texture properties observed above can be attributed to several factors. As the pressure increases, the casein micelles become smaller, which leads to an increase of effective surfaces with interaction forces and the formation of a strong gel network in skimmed yogurt [32]. Additionally, the temperature of raw milk at the outlet increased with the increase of pressure in the process of using the UHPJ (Figure 1). Under the combined effects of heating and pressure, the denaturation of the whey protein (β-lactoglobulin, immunoglobulins, and α-lactalbumin) intensified [28–30] and the solubility of the protein decreased, which also accelerated the coagulation of milk. Furthermore, the UHPJ process induced the formation of covalent disulfide bonds between casein and whey proteins, which contributed to the formation of a more compact gel and increased its water retention [33].

3. Materials and Methods

3.1. Materials and Reagents

For this study, fresh bovine milk was provided by the farm passing the GAP first-level certification (Tianjin, China). The chemical composition of raw milk is shown in Table 1. YoFlex® Premium 1.0, the direct VAT inoculation composed of *Lactobacillus bulgaricus* and *Streptococcus thermophilus*, was provided by Chr. Hansen Holding A/S (Hoersholm, Denmark). Tris, glycine, ethylene diamine tetraacetic Acid (EDTA), dinitrobenzoic acid (DTNB), trichloroacetic acid (TCA), urea, and β-mercaptoethanol were all purchased from Beijing Solarbio Science & Technology Co., Ltd. All other chemicals and reagents used were of analytical grade.

3.2. Degreasing

The preparation of the sample and the subsequent experimental process are shown in Figure 8. A pilot-type milk fat separator (GEA Group Co., Ltd., Bochum, Germany) was used for cold degreasing treatment of the fresh milk. The treatment flow was 450 L/h with a rotational speed of 9000–10,100 r/min.

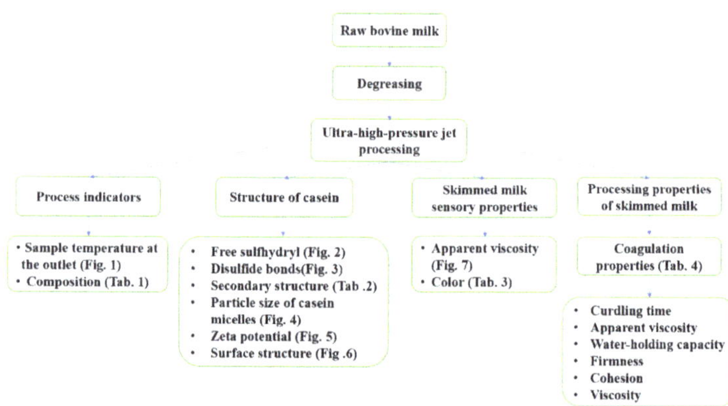

Figure 8. Flow chart of the experiment.

3.3. UHPJ Processing

The milk was treated using ultra-high pressures of 100, 150, 200, 250 and 300 MPa (Dynamic ultra-high-pressure jet processing homogenizer, laboratory's self-developed equipment, Tianjin, China) immediately after skimming. Each sample was processed only once. Samples were collected and cooled to 8 °C using a mixture of ice and water; then the samples were stored and refrigerated at 4 °C.

3.4. Analysis of Milk Composition

The composition of milk was determined by Milko Scan FT-120 (FOSS Analytical A/S, Denmark). Sample injection, testing and data analysis were controlled by the FT-120 software program, with an input temperature of 45 °C and injection volume of 30 mL.

3.5. Extraction of Bovine Casein

The skimmed milk treated by UHPJ processing was cooled to room temperature and its pH value was adjusted to 4.6 with a 1 mol/L HCl solution. Next, it was centrifuged at 4 °C and 7000 r/min for 15 min (high-speed freezing centrifuge 3K15, Sigma Laborzentrifugen GmbH, Osterode, Germany), after which the underlying precipitate was casein. This precipitate was washed several times with deionized water until it was neutral, then freeze-dried and preserved.

3.6. Determination of Sulfhydryl and Disulfide Bonds in Casein

The contents of the free sulfhydryl group and the disulfide bond of casein were determined according to Wang C.Y.'s [10] method.

For the determination of the free sulfhydryl content (SH_F), 1 mL of casein solution with a concentration of 10 mg/mL was mixed with 5 mL of Tris-Gly-8M urea solution (0.086 mol/L Tris, 0.09 mol/L glycine, 0.004 mol/L EDTA and 8 mol/L urea) and 0.04 mL of DTNB solution (4 mg/mL), and then allowed to react at 25 °C for 30 min. The samples were monitored by measuring absorbance at 412 nm. The determination of the blank control was carried out using distilled water instead of casein.

For the determination of total sulfhydryl content (SH_T), 1 mL of casein solution with a concentration of 10 mg/mL was mixed with 5 mL of Tris-Gly-10 M urea solution (0.086 mol/L Tris, 0.09 mol/L glycine, 0.004 mol/L EDTA and 10 mol/L urea) and 0.1 mL of β-mercaptoethanol, and then allowed to react at 25 °C for 1 h. Next, 10 mL of 12% TCA solution was added and this mixture was maintained at the same temperature for 1 h, then centrifugated at 3000 r/min for 10 min. After that, the precipitate was washed with 12% TCA solution and centrifuged for 15 min at 4000 r/min. This was repeated three times. The samples were monitored by measuring absorption at 412 nm, and distilled water was used for the control.

Sulfhydryl groups and disulfide bonds were calculated as follows:

$$\text{Sulfhydryl group content } (\mu\text{mol/g protein}) = \frac{73.53 \times A_{412}}{C} \quad (1)$$

$$\text{Disulfide bond content } (\mu\text{mol/gprotein}) = \frac{SH_T - SH_F}{2} \quad (2)$$

A_{412} was the absorbance at λ = 412 nm; C was the casein concentration of the sample in mg/mL.

3.7. Analysis of the Secondary Structure of Casein by Circular Dichroism

The secondary structure of casein was determined by circular dichroism spectrometer (MOS-500, Bio-Logic Science Instruments, Seyssinet-Pariset, France). The lyophilized casein powder was dissolved in a 0.02 mol/L phosphate buffer (pH 7.5) at a concentration of 0.1 mg/mL. Detection was carried out in the range of 190–260 nm with a step of 1 nm; one data point was collected every second. The same phosphate buffer was used as a blank. The secondary structure of each sample, including α-helix, β-fold, β-turning and random coil, was fitted and calculated by CDNN software to obtain the relative percentage content.

3.8. Determination of Particle Size and Zeta Potential of Casein

The particle size and Zeta potential of casein micelles in skimmed milk were determined by a laser nano-particle size potentiometer (Zetasizer Nano ZS, Malvern Panalytical, Malvern, UK). The skimmed milk was diluted 100 times with ultrapure water. The particle size of the diluent was measured at a diffraction angle of 173°. The test mode was a protein and the solution mode water. The Zeta potential of the same diluted skimmed milk was measured at room temperature. The test mode was automatic.

3.9. Surface Structure of Casein Mapped by Scanning Electron Microscope (SEM)

Following the method of Zhang A.Q. et al. [34], the microstructure of casein was observed with an SU8100 scanning electron microscope (Hitachi, Ltd., Tokyo, Japan). The freeze-dried casein powder was placed in a 1.5 mL centrifuge tube and fixed with glutaraldehyde for more than 4 h. Samples were washed 3 times at 10 min per time with a 0.1 mol/L phosphate buffer and were then fixed with 1% osmium tetroxide for 1 h.

The samples were gradient dehydrated with 50%, 70%, 80%, 90% and 100% ethanol for 10 min each time. The centrifugal tube was dried in a dryer for 12 h. Subsequently, the sample was taken out and adhered to a copper plate for conductive treatment. The sample was then observed by scanning electron microscope and the acceleration voltage was found to be 3 kV.

3.10. Measurement of Color of Skimmed Milk

Following the method employed by Chi X.L. et al. [35], a colorimeter (TS7700, Shenzhen ThreeNH Technology Co., Ltd., Shenzhen, China) was used to assess the color of skimmed milk after UHPJ processing. The Commission Internationale de l'Eclairage (CIE) L^*, a^* and b^* of samples were measured with an illuminant of D65 at a standard 10°. ΔL^*, Δa^* and Δb^*, which represent the differences between L^*, a^* and b^* of each sample and the control, respectively, were calculated and subsequently the color difference value of each sample was calculated according to the following formula:

$$\Delta E* = \sqrt{(\Delta L^*)^2 + (\Delta a^*)^2 + (\Delta b^*)^2} \tag{3}$$

3.11. Fermentation of Skimmed Milk

Skimmed milk treated by an UHPJ was inoculated with starter (YoFlex® Premium 1.0, 0.02 U per 50 g of skimmed milk), shaken well, and fermented in a water bath at 42 °C until its pH reached 4.6. The curdling time of the sample was recorded. After the curd structure was destroyed by hand stirring, the curd was refrigerated at 4 °C for more than 4 h.

3.12. Properties of Skimmed Milk and Curd Texture

The apparent viscosity of skimmed milk and that of curd was measured with a rheometer (MCR302, Anton Paar, Graz, Austria). Then a texture analyzer was used to determine the curd's hardness, cohesiveness, and viscosity (TA-XT Plus, Stable Systems Co., Ltd., U.K.). Texture measurement conditions were as follows: probe: A/BE35; Test distance: 25.0 mm; Probe speed: 2.0 mm/s; Inductive force: Auto-10.0 g. Each of the samples was measured 3 times.

3.13. Water-Holding Capacity of Curd

An amount of 25 g of the curd sample was weighed and recorded as W_1. After being centrifuged at 4500 r/min (3K15, Sigma Laborzentrifugen GmbH, Osterode, Germany) for 15 min, the curd rested for 10 min. The supernatant was removed, after which the weight of the lower precipitate was measured and recorded as W_2.

The water holding capacity of curd was calculated according to the following formula:

$$\text{water holding capacity (\%)} = (W_2/W_1) \times 100\% \tag{4}$$

3.14. Data Processing and Analysis

The experiment was conducted three times to obtain all data, the outcomes of which were expressed as means ± SD. The significance threshold for the statistical analysis was set to 0.05 using SPSS 19.0.

4. Conclusions

The results of this study show that UHPJ processing not only alters the secondary and tertiary structures of casein, but also reduces the average particle size and Zeta potential of casein in skimmed milk, while also inducing changes in the microstructure of casein. The combination of these changes affects sensory properties, such as the color and viscosity of skimmed milk. At the same time, the curdling time of skimmed milk was effectively shortened, and the coagulation properties, such as the water holding capacity and the viscosity were significantly increased. As a result, UHPJ processing, which combines homogenization and sterilization functions, has prominent advantages for the raw milk pretreatment of yogurt, and these can be applied in dairy manufacture to improve the fermentation efficiency and the texture of fat-free yogurt.

However, the change in the coagulation property is the result of the combination of changes in both the casein and the whey proteins. Further research, which is currently being carried out by our team, will focus on the evaluation of the effects of the UHPJ processing on whey protein.

Author Contributions: Conceptualization, L.X. and Y.W.; methodology, F.X., P.Z. and Z.W.; data curation, F.X. and Y.M.; writing—original draft preparation, F.X. and Y.M.; writing—review and editing, L.X.; supervision, T.N.; funding acquisition, Z.W., P.Z. and Y.W. All authors have read and agreed to the published version of the manuscript.

Funding: This research was funded by Tianjin Key Research and Development Plan (22YFFCYS00060), Central-Guided Local Science and Technology Development Fund Projects (22ZYCGSN00110), and Beijing Postdoctoral Research Foundation (2022-ZZ-121).

Institutional Review Board Statement: Not applicable.

Informed Consent Statement: Not applicable.

Data Availability Statement: Data are contained within the article.

Conflicts of Interest: The authors declare no conflict of interest.

References

1. Liu, W.; Liu, J.H.; Xie, M.Y.; Liu, C.M.; Liu, W.L.; Wan, J. Characterization and High-Pressure Microfluidization-Induced Activation of Polyphenoloxidase from Chinese Pear (*Pyrus pyrifolia* Nakai). *J. Agric. Food Chem.* **2009**, *57*, 5376–5380. [CrossRef]
2. Liu, G.X.; Tu, Z.C.; Wang, H.; Zhang, L.; Huang, T.; Ma, D. Monitoring of the functional properties and unfolding change of ovalbumin after DHPM treatment by HDX and FTICR MS: Functionality and unfolding of oval after DHPM by HDX and FTICR MS. *Food Chem.* **2017**, *227*, 413–421. [CrossRef]
3. Shen, L.; Tang, C. Microfluidization as a potential technique to modify surface properties of soy protein isolate. *Food Res. Int.* **2012**, *48*, 108–118. [CrossRef]
4. Gong, K.J.; Chen, L.R.; Xia, H.Y. Driving forces of disaggregation and reaggregation of peanut protein isolates in aqueous dispersion induced by high-pressure microfluidization. *Int. J. Biol Macromol.* **2019**, *130*, 915–921. [CrossRef]
5. Zhang, Z.Y.; Yang, Y.L.; Zhou, P.; Zhang, X.; Wang, J.Y. Effects of high pressure modification on conformation and gelation properties of myofibrillar protein. *Food Chem.* **2017**, *217*, 678–686. [CrossRef]
6. Fan, Q.Y.; Wang, P.X.; Zheng, X.Y.; Hu, J.M. Effect of dynamic high pressure microfluidization on the solubility properties and structure profiles of proteins in water-insoluble fraction of edible bird's nests. *LWT-Food Sci. Technol.* **2020**, *132*, 109923. [CrossRef]
7. Ginger, M.R.; Grigor, M.R. Comparative aspects of milk caseins. *Comp. Biochem. Phys. B* **1999**, *124*, 133–145. [CrossRef]
8. Kumar, A.; Badgujar, P.C.; Mishra, V.; Upadhyay, A. Effect of microfluidization on cholesterol, thermal properties and in vitro and in vivo protein digestibility of milk. *LWT-Food Sci. Technol.* **2019**, *116*, 108523. [CrossRef]
9. Tran, M.; Roberts, R.; Felix, T.L.; Harte, F.M. Effect of high-pressure-jet processing on the viscosity and foaming properties of pasteurized whole milk. *J. Dairy Sci.* **2018**, *101*, 3887–3899. [CrossRef]
10. Bucci, A.J.; Van Hekken, D.L.; Tunick, M.H.; Tomasula, P.M. The effects of microfluidization on the physical, microbial, chemical, and coagulation properties of milk. *J. Dairy Sci.* **2018**, *101*, 6990–7001. [CrossRef]
11. Pereda, J.; Ferragut, V.; Quevedo, J.M.; Guamis, B.; Trujillo, A.J. Effects of Ultra-High Pressure Homogenization on Microbial and Physicochemical Shelf Life of Milk. *J. Dairy Sci.* **2007**, *90*, 1081–1093. [CrossRef]
12. Xing, B.B.; Zhang, T.T.; Zhao, Q.; Xiong, H. Effect of High Pressure Microfluidization Treatment on the Properties of Thermal Glutelin Aggregates. *Food Sci.* **2019**, *40*, 109–115.
13. Wang, C.Y.; Ma, Y.P.; Liu, B.G.; Kang, Z.L.; Geng, S.; Wang, J.N.; Wei, L.P.; Ma, H.J. Effects of dynamic ultra-high pressure homogenization on the structure and functional properties of casein. *Int. J. Agric. Biol. Eng.* **2019**, *12*, 229–234. [CrossRef]
14. De Kruif, C.G.; Holt, C. Casein Micelle Structure, Functions and Interactions. In *Advanced Dairy Chemistry-1 Proteins*, 3rd ed.; Fox, P.F., McSweeney, P.L.H., Eds.; Springer: Boston, MA, USA, 2003; pp. 233–275.
15. Farrell, H.M.; Jimenez-Flores, R.; Bleck, G.T.; Swaisgood, H.E. Nomenclature of the Proteins of Cows' Milk—Sixth Revision. *J. Dairy Sci.* **2004**, *87*, 1641–1674. [CrossRef]
16. Daiki, O.; Wataru, O.; Shojiro, T.; Tomohiro, N.G.; Takano, K. Effects of the thermal denaturation degree of a whey protein isolate on the strength of acid milk gels and the dissociation of κ-casein. *J. Dairy Res.* **2022**, *89*, 104–108.
17. Zhou, X.F.; Zheng, Y.R.; Zhong, Y.; Wang, D.F.; Deng, Y. The Effect of DPCD on the Structure and Physical Properties of Casein. *J. Chin. Inst. Food Sci. Technol.* **2022**, *22*, 120–128.

18. Mohan, M.S.; Ye, R.; Harte, F. Initial study on high pressure jet processing using a modified waterjet on physicochemical and rennet coagulation properties of pasteurized skim milk. *Int. Dairy J.* **2016**, *55*, 52–58. [CrossRef]
19. López-Fandiño, R. High pressure-induced changes in milk proteins and possible applications in dairy technology. *Int. Dairy J.* **2005**, *16*, 1119–1131. [CrossRef]
20. Roach, A.; Harte, F. Disruption and sedimentation of casein micelles and casein micelle isolates under high-pressure homogenization. *Innov. Food Sci. Emerg. Technol.* **2008**, *9*, 1–8. [CrossRef]
21. Lunardi, C.N.; Gomes, A.J.; Rocha, F.S.; De Tommaso, J.; Patience, G.S. Experimental methods in chemical engineering: Zeta potential. *Can. J. Chem. Eng.* **2021**, *99*, 627–639. [CrossRef]
22. Janahar, J.J.; Marciniak, A.; Balasubramaniam, V.M.; Jimenez-Flores, R.; Ting, E. Effects of pressure, shear, temperature, and their interactions on selected milk quality attributes. *J. Dairy Sci.* **2021**, *104*, 1531–1547. [CrossRef]
23. Sandra, S.; Dalgleish, D.G. Effects of ultra-high-pressure homogenization and heating on structural properties of casein micelles in reconstituted skim milk powder. *Int. Dairy J.* **2005**, *15*, 1095–1104. [CrossRef]
24. Hu, X.; Zhao, M.; Sun, W.; Zhao, G.; Ren, J. Effects of microfluidization treatment and transglutaminase cross-linking on physicochemical, functional, and conformational properties of peanut protein isolate. *J. Agric. Food Chem.* **2011**, *59*, 8886. [CrossRef]
25. Huppertz, T.; Fox, P.F.; Kelly, A.L. Properties of casein micelles in high pressure-treated bovine milk. *Food Chem.* **2003**, *87*, 103–110. [CrossRef]
26. Hettiarachchi, C.A.; Corzo-Martínez, M.; Mohan, M.S.; Harte, F.M. Enhanced foaming and emulsifying properties of high-pressure-jet-processed skim milk. *Int. Dairy.* **2018**, *87*, 60–66. [CrossRef]
27. Nassar, K.S.; Zhang, S.W.; Lu, J.; Pang, X.Y.; Ragab, E.S.; Yue, Y.C.; Lv, J.P. Combined effects of high-pressure treatment and storage temperature on the physicochemical properties of caprine milk. *Int. Dairy J.* **2019**, *96*, 66–72. [CrossRef]
28. Serra, M.; Trujillo, A.J.; Jaramillo, P.D.; Guamis, B.; Ferragut, V. Ultra-High Pressure Homogenization-Induced Changes in Skim Milk: Impact on Acid Coagulation Properties. *J. Dairy Res.* **2008**, *75*, 69–75. [CrossRef]
29. Ravash, N.; Peighambardoust, S.H.; Soltanzadeh, M.; Pateiro, M.; Lorenzo, J.M. Impact of high-pressure treatment on casein micelles, whey proteins, fat globules and enzymes activity in dairy products: A review. *Crit. Rev. Food Sci.* **2022**, *62*, 2888–2908. [CrossRef]
30. Panagiotis, S.F.; Constatnina, T. Conventional and Innovative Processing of Milk for Yogurt Manufacture; Development of Texture and Flavor: A Review. *Foods.* **2014**, *3*, 176–193.
31. Hernandez, A.; Harte, F.M. Manufacture of Acid Gels from Skim Milk Using High-Pressure Homogenization. *J. Dairy Sci.* **2008**, *91*, 3761–3767. [CrossRef]
32. Ciron, C.I.E.; Gee, V.L.; Kelly, A.L.; Auty, M.A.E. Comparison of the effects of high-pressure microfluidization and conventional homogenization of milk on particle size, water retention and texture of non-fat and low-fat yoghurts. *Int. Dairy J.* **2010**, *20*, 314–320. [CrossRef]
33. Morin, P.; Jiménez-Flores, R.; Pouliot, Y. Effect of processing on the composition and microstructure of buttermilk and its milk fat globule membranes. *Int. Dairy J.* **2007**, *17*, 1179–1187. [CrossRef]
34. Zhang, A.Q.; Wang, Y.Y.; Li, R.; Zhou, G.W.; Wang, L.; Wang, X.B. Effect of Ultra-Pasteurization on Microstructure and Aggregation Properties of Casein from Bovine Milk. *Food Sci.* **2020**, *41*, 106–110.
35. Chi, X.L.; Tong, L.J.; Pan, M.H.; Alifu, N.; Ai, N.S.; Wang, B.; Sun, B.G. Effect of Fat Content on Physicochemical Properties of Milk. *Food Sci.* **2018**, *39*, 26–31.

molecules MDPI

Article

Effect of Steam Flash-Explosion on Physicochemical Properties and Structure of High-Temperature Denatured Defatted Rice Bran Protein Isolate

Zhiguo Na [1], Haixin Bi [1,2], Yingbin Wang [2], Yujuan Guo [2] and Yongqiang Ma [1,*]

1 School of Food Engineering, Harbin University of Commerce, Harbin 150028, China
2 College of Food Engineering, East University of Heilongjiang, Harbin 150060, China
* Correspondence: mayq@hrbcu.edu.cn

Abstract: The effects of Steam Flash-Explosion (SFE) on the physicochemical properties and molecular structure of high-temperature denatured defatted rice bran protein isolate (RBPI) were investigated. The mechanism of SFE treatment on high-temperature denatured defatted RBPI was revealed. The analysis of the physical and chemical properties of RBPI showed that the surface hydrophobicity, characteristic viscosity, and thermal stability of rice bran protein isolate were significantly affected by the pressure of saturated steam and pressure holding time. Under the conditions of 2.1 MPa and 210 s, the surface hydrophobicity index decreased significantly from 137.5 to 17.5, and the characteristic viscosity increased significantly. The peak temperature of denaturation decreases from 114.2 to 106.7 °C, and the enthalpy of denaturation decreases from 356.3 to 231.4 J/g. The higher structure (circular dichroic spectrum and endogenous fluorescence spectrum) of rice bran protein isolate was analyzed by volume rejection chromatography (SEC). The results showed that steam flash treatment could depolymerize and aggregate RBPI, and the relative molecular weight distribution changed greatly. The decrease in small molecules with poor solubility was accompanied by the increase in macromolecules (>550 kDa) soluble aggregates, which were the products of a Maillard reaction. The contents of free sulfhydryl and disulfide bonds in high-temperature rice bran meal protein isolate were significantly increased, which resulted in the increase in soluble aggregates containing disulfide bonds. Circular dichroism (CD) analysis showed that the α-helix content of the isolated protein was significantly decreased, the random curl content was increased, and the secondary structure of the isolated protein changed from order to disorder. The results of endogenous fluorescence spectroscopy showed that the high-temperature rice bran meal protein isolate was more extended, tryptophan was in a more hydrophilic microenvironment, the fluorescence intensity was reduced, and the tertiary structure was changed. In addition, the mean particle size and net surface charge of protein isolate increased in the aqueous solution, which was conducive to the development of the functional properties of the protein.

Keywords: high-temperature denatured defatted rice bran; steam flash-explosion; rice bran protein isolate; structural properties

Citation: Na, Z.; Bi, H.; Wang, Y.; Guo, Y.; Ma, Y. Effect of Steam Flash-Explosion on Physicochemical Properties and Structure of High-Temperature Denatured Defatted Rice Bran Protein Isolate. *Molecules* **2023**, *28*, 643. https://doi.org/10.3390/molecules28020643

Academic Editors: Shudong He, Wei Xu and Muhammad H. Aludatt

Received: 8 November 2022
Revised: 26 December 2022
Accepted: 4 January 2023
Published: 8 January 2023

1. Introduction

Rice bran, as the main by-product of rice production, is extremely rich in resources. It not only has a huge yield and low price but also concentrates more than 60% of the nutrients of rice [1], has a high content of unsaturated fatty acids such as linoleic acid, up to 80%, and is rich in dozens of natural bioactive substances such as vitamin E, tocopherol trienes and squalene [2]. Compared with other plant proteins, the lysine content in rice bran protein is higher, which not only requires complete amino acids, reasonable composition, close to the FAO/WHO recommended model, and high biological titer but also has the advantages of being hypoallergenic, which is a plant protein resource with great potential for development [3]. The protein content is about 11.3–14.9%. Soluble protein accounted for about 70% of the total, containing all four proteins in Osbron's classification, namely albumin, globulin, gluten, and glycolysin [4], with a nutritional value comparable to that of egg protein [5]. However, these proteins have strong polymerization and a large number of disulfide bonds, and the high content of dietary fiber and phytic acid in rice bran aggravates the intercoalescence of rice bran components and the functional properties such as solubility decline [6], which makes the extraction and separation of proteins difficult and greatly limits its application in the food industry.

The Steam flash-explosion (SFE) treatment can cause the material to be subjected to the mechanical shear caused by high-temperature cooking and steam release at the same time, which can not only damage the cell wall structure of rice bran but also degrade hemicellulose and release the protein embedded in it. In addition, these effects can also cause changes in molecular structure and physical and chemical properties of proteins in high-temperature denatured defatted rice bran (HTDDRB), thus affecting functional properties such as solubility [7] and effectively improving the extraction rate and nitrogen solubility index of proteins in HTDDRB. In addition, according to the principle and characteristics of the Maillard reaction, high-temperature cooking in the steam flash process may cause covalent binding of proteins and sugars in high-temperature rice bran meal, resulting in increased polarity of protein molecules and changes in spatial structure, leading to changes in solubility [8]. SFE technology, as a typical physicochemical treatment technology, is a treatment method in which raw materials are placed in a closed environment of high-temperature liquid water and high-pressure steam for a certain period of time and then immediately released to atmospheric pressure. It is economical, efficient, and pollution-free [9,10]. At present, there are few reports on the effect of steam flash detonation treatment on the physicochemical properties and structure of rice bran protein isolate.

Therefore, on the basis of previous studies, taking high-temperature rice bran meal protein isolates with high nitrogen dissolution index and high protein extraction rate as the research object, this study analyzed the effects of SFE treatment on the physicochemical properties and structure of high-temperature denatured defatted rice bran protein isolate, and discussed its mechanism of action, providing a theoretical basis for the development and utilization of high-temperature rice bran protein.

2. Results and Discussion

2.1. Surface Hydrophobicity (H_0) Analysis

Surface hydrophobicity is an important physical and chemical property of proteins, which is of great significance for protein stability, conformation, and function [11]. The interaction of proteins with water and other compounds in solution is closely related to surface hydrophobicity, so the solubility, emulsification, water absorption, and other functional properties of proteins are affected by surface hydrophobicity. The surface hydrophobicity is also affected by the higher structure of proteins. For example, when proteins are subjected to high-temperature, high-pressure, shear force, and other factors, the higher structure will change, thus affecting the surface hydrophobicity of proteins [12]. ANS fluorescent probe method is the most commonly used method for the determination of protein surface hydrophobicity because it has the advantages of simple operation, fast operation, and a small amount of protein. As a fluorescence probe sensitive to polarity, 8-aniline-1-naphthalene sulfonic acid (ANS) can combine with the hydrophobic cavity of protein molecules to enhance the fluorescence intensity of proteins, and the fluorescence intensity will increase with the addition of ANS until excessive fluorescence probe is added, so it can be used to determine the surface hydrophobicity of proteins [13]. Figure 1 shows the surface hydrophobicity of rice bran protein isolates prepared by high-temperature rice bran meal treated by steam flash under different conditions.

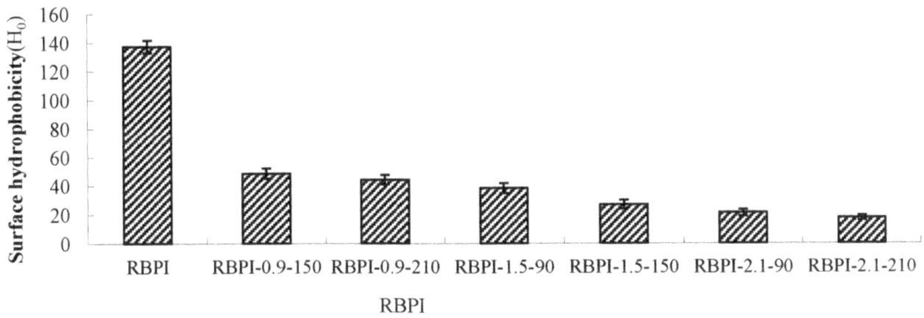

Figure 1. Surface hydrophobicity of RBPI prepared from different SFE pretreatment conditions.

As can be seen from Figure 1, the surface hydrophobicity of RBPI decreased significantly ($p < 0.05$) after SFE treatment, and the maximum decrease was 87.3%, from 137.5 (RBPI) to 17.5 (rbPI-2.1-210), and the surface hydrophobicity of protein isolates was greatly affected by steam pressure and pressure holding time. This is consistent with the research results of Zhang Yanpeng et al. [14]. The decrease in surface hydrophobicity indicated that the structure and physicochemical properties of RBPI changed obviously during SFE. The RBPI of HTDDRB can be subjected to various physical and chemical effects, such as high-temperature cooking, mechanical shear force, and Maillard reaction caused by the instant release of steam. Among them, mechanical shearing can depolymerize protein aggregates, and some hydrophobic groups leak out. With the extension of pressure holding time, high-temperature cooking is strengthened, and these proteins gather again due to hydrophobic action, and the hydrophobic groups leaked out are hidden inside molecules, reducing the hydrophobic surface property of proteins [15]. In addition, previous studies have confirmed that the Maillard glycosylation reaction triggered by SFE can also significantly affect the surface hydrophobicity of RBPI [16]. Achouri et al. [17] modified 11S globulin by glycation and found that the decrease in surface hydrophobicity was related to the improvement of grafting degree. Meanwhile, glycation increased the hydroxyl group on the protein molecule, thus increasing the hydrophilicity of the protein, and some hydrophobic groups were shielded. Due to the Maillard reaction, the free amino group on the amino acid side chain of the protein binds sugar molecules, which affects the binding of protein and ANS, thus reducing the measured value of surface hydrophobicity. Gasymov et al. [18] found that protein glycosylation reduced lysine and arg in protein, thus reducing the binding site of ANS and protein. This is consistent with the results of this study.

2.2. Intrinsic Viscosity Analysis

The intrinsic viscosity [η] is a physical property of polymer solution. Polymer solution concentration tends to zero specific viscosity, that is, polymer solution in infinite dilution of a single polymer molecule and the internal friction between solvent molecules; because the solution is very thin, polymer molecules are far away from each other, so the intrinsic viscosity [η] value is not affected by the solution concentration. However, [η] values are related to the molecular weight and molecular morphology of the polymer and increase with the increase in molecular weight and molecular stretching degree of the polymer.

Figure 2 shows the change in Intrinsic viscosity of RBPI from HTDDRB after SFE. As shown in Figure 2, the Intrinsic viscosity of RBPI prepared by HTDDRB after SFE increased significantly ($p < 0.05$), from 28.6 to 68.7 mL/g by 140.2%, and showed a trend of increasing with the increase in steam pressure and pressure holding time. The results indicated that the stretch degree of RBPI prepared by SFE was significantly higher than that of untreated RBPI, and the internal friction between protein and solvent molecules was increased, which was greatly affected by steam pressure and pressure holding time. This is mainly because of the process of SFE. The material is first subjected to heat treatment and the mechanical shear brought by the rapid movement of high-pressure steam. The heat treatment can make the protein partially denature and allow molecular aggregation to occur.

At the same time, the mechanical shear brought by high-pressure steam can destroy the aggregation state of proteins and cause molecular stretching. The force is greatly affected by steam pressure. Therefore, the increase in steam pressure leads to an increase in Intrinsic viscosity and molecular stretching. The extension of protein molecules exposed more free amino groups, promoted the occurrence of the Maillard reaction, introduced more sugar molecules, and changed the spatial structure of proteins. Therefore, the extension of the pressure holding time is conducive to the occurrence of the Maillard reaction, which makes protein molecules more extended. In the later stage of steam flash, the material is subjected to shear action caused by the instantaneous release of high-pressure steam, which further causes the depolymerization of protein molecules, and the molecules are more extended. Thus, the Intrinsic viscosity is higher.

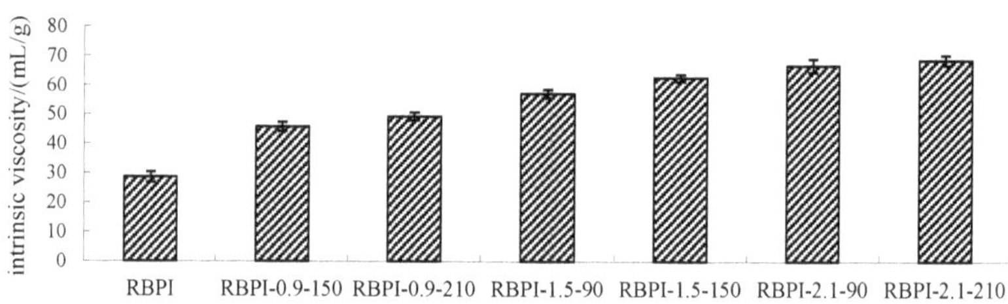

Figure 2. Intrinsic viscosity of RBPI prepared from different SFE pretreatment conditions.

2.3. Thermal Stability (DSC) Analysis

Differential scanning calorimetry (DSC) is widely used to study the thermal stability of proteins. When heated, the hydrogen bonds inside proteins break, causing the protein molecules to stretch, a process that requires energy absorption, known as denaturation heat. Protein denaturation is manifested by molecular structure changing from folded state to unfolded state, from an ordered state to a disordered state, and from a natural state to a denatured state, and these processes are accompanied by energy changes. Thermal analysis of proteins involves the destruction of higher structures by heat and the measurement of energy changes in the process.

Figure 3 and Table 1, respectively, show the DSC curve and thermodynamic characteristics of RBPI before and after SFE (steam pressure 2.1 MPa, holding time 210 s). It can be seen that the denaturation peak of RBPI before and after steam flash detonation is upward, indicating that the RBPI absorbs heat during the denaturation process. The denaturation peak temperatures and enthalpy changes of original RBPI and RBPI-2.1-210 were 114.2 °C and 106.7 °C, respectively, and the enthalpy changes of denaturation were 356.3 J/g and 231.4 J/g, respectively. The denaturation temperatures and enthalpy changes of the two were significantly different, indicating that the structure and composition of the two were significantly different.

Table 1. Thermal transition characteristics of rice bran protein isolate.

Protein Sample	Starting Temperature T_o/°C	Peak Temperature T_p/°C	End Temperature T_e/°C	Enthalpy ΔH (J/g)
RBPI	77.8	114.2	179.2	356.3
RBPI-2.1-210	80.7	106.7	142.6	231.4

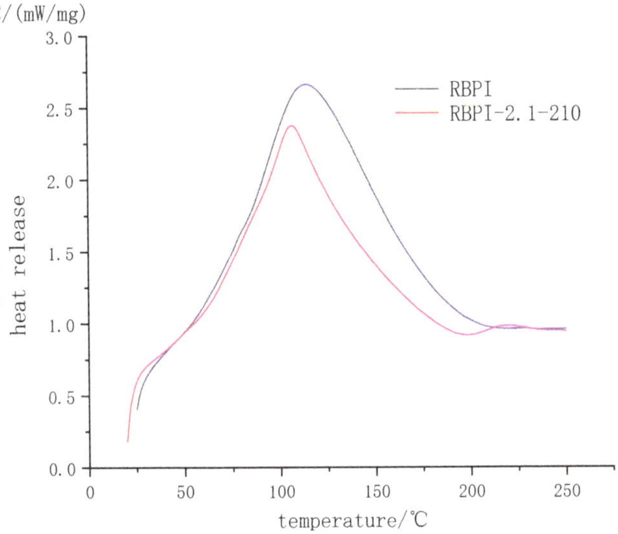

Figure 3. DSC Profile of rice bran protein isolate.

Non-covalent bonds such as hydrogen bonds and hydrophobic interactions play an important role in the stability of protein molecular structure. They may be reversed bonds, which play a major role in thermal denaturation temperature and enthalpy change. The thermal denaturation temperature of plant proteins is closely related to their secondary structure and is generally determined by non-covalent bonds [19]. In this study, the denaturation temperature of RBPI-2.1-210 decreased after SFE, indicating that the thermal stability of the protein decreased, while the decrease in enthalpy change value indicated that the molecular structure of the protein was locally expanded, the hydrophobic groups were closer to the polar microenvironment, and some internal hydrogen bonds were likely to break, and the protein molecules were closer to the disordered state. Although raw materials are subjected to high-temperature cooking in the process of the steam flash explosion, the thermal denaturation is not obvious due to the very short time of action. Meanwhile, proteins are also subjected to mechanical shear caused by the instantaneous release of steam, which can damage non-covalent bonds in proteins, partially break hydrogen bonds, and reduce surface hydrophobicity (previous studies have confirmed this). At the same time, due to the introduction of sugar molecules into RBPI by the Maillard reaction, the non-covalent bond of protein molecules was damaged, which resulted in a decrease in thermal denaturation temperature. In addition, the thermal stability of proteins is also closely related to disulfide bonds in protein molecules. More disulfide bonds in protein molecules can promote a more stable conformation of proteins [20]. Kinsella et al. [21] found that the more disulfide bonds there are, the higher the thermal stability of proteins is. It can be inferred that steam flash treatment may damage the disulfide bond of RBPI to some extent and reduce thermal stability.

2.4. Size Exclusion Chromatography (SEC) Analysis (Relative Molecular Mass Distribution)

Figure 4 shows the effect of SFE treatment conditions on the relative molecular weight distribution and aggregation state of RBPI. The main peak locations of RBPI prepared before and after steam flash treatments were 8.27 min, 15.24 min, 17.28 min, 22.8 min, and 23.5 min, and their corresponding relative molecular weights were 654.4 kDa, 350.1 kDa, 238 kDa, respectively. The peaks with 24.9 kDa, 9.8 kDa molecular weight, and an 8.27 min retention time, correspond to the protein molecular aggregates (654.4 kDa). As can be seen from Figure 4, the peak value of protein isolate (RBPI) from rice bran meal at 8.27 min is very small, and the peak value of RBPI increases significantly after steam flash treatment, while the peak value of RBPI at other locations decreases. The results indicated that there were few macromolecular weight aggregates in the RBPI, but the macromolecular weight aggregates were formed in the RBPI after SFE.

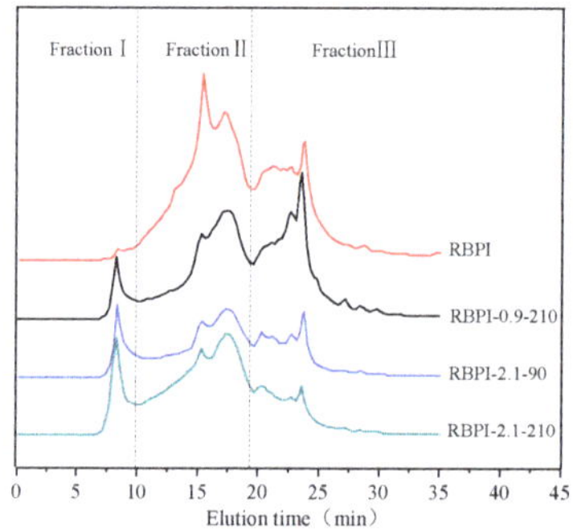

Figure 4. SEC profiles of RBPI prepared from different SFE pretreatment conditions.

In order to further analyze the relative molecular mass distribution changes of rice bran protein isolate, the chromatogram was fitted using Peakfit v4.12, and the peak integral area of each part was calculated. According to the molecular weight distribution of rice bran protein isolate, it was divided into FractionI (>550 kDa). For FractionII (550~80 kDa) and Fraction III (<80 kDa), the effect of steam blitzing treatment on rice bran protein aggregates can be reflected through the change in peak integral area of each elution region, as shown in Table 2.

Table 2. Molecular weight distribution of RBPI prepared from different SFE pretreatment conditions (%).

Elution Peak Area	RBPI	RBPI-0.9-210	RBPI-2.1-90	RBPI-2.1-210
Fraction I (>550 kDa)	1.5 ± 0.08	6.7 ± 0.34	9.3 ± 0.16	13.8 ± 0.49
Fraction II (550~80 kDa)	58.1 ± 1.02	43.9 ± 0.51	44.5 ± 1.03	50.8 ± 0.83
Fraction III (<80 kDa)	40.4 ± 0.53	49.4 ± 0.24	46.2 ± 0.68	35.4 ± 0.15

It can be seen from Table 2 that the relative content of Fraction I of isolated protein increased from 1.5 to 13.8% after SFE treatment, and the relative content of Fraction II and Fraction III decreased significantly, indicating that steam flash burst can promote the formation of soluble protein aggregates [22]. In addition, it can be seen from Table 2 that the distribution of the relative molecular weight of rice bran protein is greatly affected by the steam pressure and holding time, and Fraction II decreases from 58.1 to 43.9% at lower steam pressure (0.9 MPa), accompanied by the increase in Fraction I and Fraction III., indicating that some rice bran proteins depolymerize under mechanical shearing at lower pressure, and molecular aggregation occurs under high-temperature cooking, and with the increase in steam pressure, The relative content of small molecular weight Fraction III decreased significantly ($p < 0.05$), from 49.4 to 35.4%, and with the increase in Fraction I and Fraction III., the effect of holding time on relative molecular mass was basically the same as that of vapor pressure, indicating that higher vapor pressure and longer holding time contributed to the unfolding of proteins to form soluble protein aggregates [23]. Since the surface properties of proteins are greatly affected by their molecular aggregation state, SFE treatment can affect the surface properties of the RBPI and then affect its functional properties.

2.5. *Mercaptol and Disulfide Bond Content*

Disulfide bonds (SS) are important covalent chemical bonds that constitute the higher structure of proteins, which play an extremely important role in stabilizing the higher structure of proteins and maintaining their active functions, and disulfide bonds (SS) and sulfhydryl groups (-SH) can be converted to each other by redox [24]. Table 3 shows the contents of free sulfhydryl (SHF) and disulfide bond (SS) of different RBPI, among which the SHF and SS of the original high-temperature rice bran meal protein isolate (RBPI) were 4.38 μmol/g and 6.13 μmol/g, respectively, while the SHF of rice bran protein isolate was 5.54 μmol/g (RBPI-0.9-210), 5.63 μmol/g (RBPI-2.1-90) and 5.15 after SFE treatment under different conditions μmol/g (RBPI-2.1-210), SS 8.88 μmol/g (RBPI-0.9-210), 9.32 μmol/g (RBPI-2.1-90), and 7.46 μmol/g (RBPI-2.1-210), it can be seen that Steam Flash Explosion can significantly increase the SS and SHF of RBPI of HTDDRB, and the steam pressure and holding time have a great influence on it, and the SS and SHF are significantly increased ($p < 0.05$) under the long-term or high-pressure short-term treatment conditions, while the two are reduced to different degrees under the long-term treatment conditions of high-pressure (2.1 MPa, 210 s).

Table 3. Free sulfhydryl and disulfide bond contents (SH_F and SS) of RBPI prepared from different SFE pretreatment conditions (μmol/g).

Project	RBPI	RBPI-0.9-210	RBPI-2.1-90	RBPI-2.1-210
SH_F	4.38 ± 0.15	5.54 ± 0.08	5.63 ± 0.14	5.15 ± 0.09
SS	6.13 ± 0.16	8.88 ± 0.15	9.32 ± 0.09	7.46 ± 0.13

The SS of the protein isolate of the original RBPI was higher than that of Wu Wei et al. [25] on the determination of fresh rice bran, which was mainly due to the oxidation of some sulfhydryl groups to disulfide bonds during the thermal stabilization of rice bran, which covalently crosslinked proteins to form thermal aggregates. In the process of SFE treatment under low-pressure long-term or high-pressure short-term conditions, the molecular structure of the protein can be destroyed due to the mechanical shearing action of high-pressure steam, and the peptide chain is stretched so that the sulfhydryl group buried inside the molecule is exposed, so that the SHF content increases. At the same time, the high-temperature cooking effect during SFE treatment can induce the formation of soluble aggregates containing disulfide bonds, which increases the SS content. When the steam pressure is 2.1 MPa, the holding time is extended from 90 to 210 s, which will greatly increase the heat treatment strength, and some studies have shown that [26] the heat treatment strength is too high but will cause some disulfide bonds to break into sulfhydryl groups, and at the same time, due to the Maillard reaction to form more non-disulfide bond covalent aggregates, free sulfhydryl groups can be wrapped inside the aggregates, so the SHF content is slightly reduced.

2.6. Secondary Structure—Circular Dichroism (CD) Analysis

Protein secondary structure refers to the local spatial arrangement (folded and coiled form) of the main chain atoms in the polypeptide chain, that is, the conformation, excluding the conformation of the side chain part, mainly including α-helix, β folding, β-angle and irregular curl, and other structural forms. Circular dichroic spectroscopy is an effective means for the study of protein secondary structure. The asymmetric α-carbon atom of amino acids has optical activity, when the plane circularly polarized light passes, these optically active centers have different absorption intensities for their left and right circularly polarized light, resulting in an absorption difference, resulting in the amplitude difference of the polarized light vector, and the plane circularly polarized light becomes elliptically polarized light, which is the circular dichroism of proteins [27].

The far-ultraviolet circular dichroic spectrum (190~250 nm) mainly reflects the protein backbone conformation, which is often used in the study of protein secondary structure. According to the literature [28], the α-helical has a positive peak near 192 nm, two negative characteristic shoulder peaks at 208 nm and 222 nm, a negative peak at 216~218 nm at β-fold, a strong positive peak at 195 nm, and a positive peak near 205 nm at the β-turn. The irregular curl has a negative peak around 198 nm and a small, wide positive peak around 220 nm. Figure 5 shows the circular dichroic spectra of different rice bran protein isolates (RBPI), from which it can be seen that the peak shape of the circular bichromatogram of rice bran protein isolate obtained after Steam Flash Explosion treatment is basically unchanged, the peak position changes slightly, and the peak intensity changes greatly. There is a strong negative peak at 216~228 nm, which is the overlapping peak of α-helical and β-rotation angle, and this peak is significantly weakened after Steam Flash Explosion. The original high-temperature rice bran meal protein isolate had a negative peak at 197 nm, and the peak position shifted to 200~202 nm after Steam Flash Explosion treatment, which was closer to the irregular coiling peak position, and the intensity increased. At the same time, it was found that a weak positive peak appeared at 195 nm after the SFE, indicating the presence of β-folding.

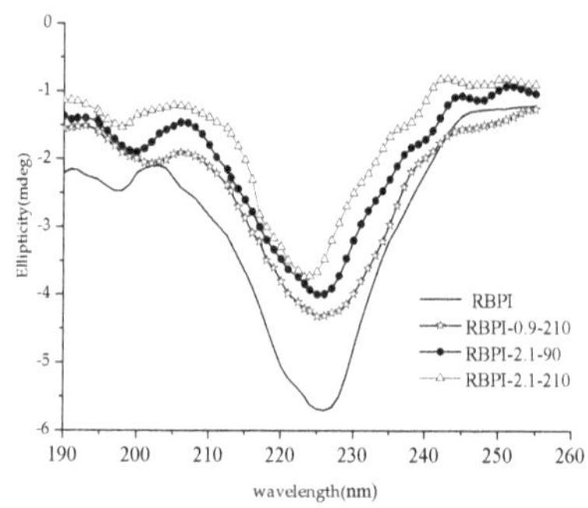

Figure 5. Circular dichroism spectra of RBPI prepared from different SFE pretreatment conditions.

The relative content of the secondary structure of RBPI is shown in Table 4. It can be seen from Table 4 that after SFE, the α-helix content of isolated protein decreased significantly ($p < 0.05$), the content of random coil increased significantly ($p < 0.05$), and the contents of β-fold and β-corner remained basically unchanged, indicating that the secondary structure changed from ordered to disordered. This is consistent with the findings of Damodarand et al. [29]. The α-helix structure plays an important role in stabilizing the protein structure, so the stability of the protein isolate molecules obtained after SFE is reduced, the molecules are fully stretched, and the flexibility is increased, which contributes to the exertion of its functional properties. This is consistent with the findings of surface hydrophobicity, thermal stability (DSC), and intrinsic viscosity and also explains these changes in secondary structure.

Table 4. Secondary structure content of RBPI prepared from different SFE pretreatment conditions (%).

Sample	α-Helix	β-Corner	Random Coil	β-Fold
RBPI	29.1 ± 0.25	34.1 ± 0.24	36.8 ± 0.36	0
RBPI-0.9-210	23.3 ± 0.18	33.2 ± 0.35	42.3 ± 0.14	1.2 ± 0.11
RBPI-2.1-90	21.5 ± 0.21	33.6 ± 0.17	43.6 ± 0.38	1.3 ± 0.09
RBPI-2.1-210	19.8 ± 0.09	32.4 ± 0.44	47.0 ± 0.23	0.8 ± 0.03

2.7. Tertiary Structure—Endogenous Fluorescence Spectroscopy

Based on the study of the thermal stability, surface hydrophobicity, and intrinsic viscosity of rice bran protein isolate, endogenous fluorescence spectroscopy was used to analyze the differences in the tertiary structure of RBPI prepared before and after SFE treatment. Taking 290 nm as the excitation wavelength, the endogenous fluorescence spectrum with tryptophan (Try) as the main emitting group can be obtained, and its peak migration can reflect the change in the polarity of the tryptophan microenvironment and then reflect the spatial conformation change in the protein (mainly characterizing the tertiary structure). The peak redshift indicates an increase in the microenvironmental polarity of the fluoroemitting group (Try) and greater exposure to solvents, while the blue shift indicates an increase in the microambient hydrophobicity of the emitting group (Try), while the decrease in fluorescence peak intensity is associated with fluorescence quenching [30]. Shutova et al. [31] found that the maximum emission peak of tryptophan residues, when exposed to the protein surface, was between 350~353 nm, and the maximum emission peak of tryptophan encapsulated in the protein was between 326~332 nm.

Figure 6 is the endogenous fluorescence spectrum of the RBPI obtained by different SFE treatments of HTDDRB, from which it can be seen that the maximum emission wavelength of the RBPI prepared from the original HTDDRB is 336 nm, and the peak position is high, indicating that the tryptophan residue of the isolated protein prepared from the original RBPI is relatively close to the hydrophilic microenvironment, and after the SFE, the maximum emission wavelength of the isolated protein undergoes different degrees of redshift, from 336 to 343, 345, 346 nm. It shows that tryptophan (Try) residues are gradually exposed on the surface of protein molecules and are in a more hydrophilic microenvironment [32], and the degree of migration is affected by the steam pressure and holding time. This is mainly caused by mechanical shearing and glycosylation during Steam Flash Explosion, which can depolymerize protein aggregates, stretch molecules, gradually expose embedded tryptophan, and covalent binding of proteins and carbohydrates can also enhance the polarity of the microenvironment [33]. At the same time, it can also be found from Figure 6 that the fluorescence intensity of RBPI is significantly reduced after a steam flash burst ($p < 0.05$), and it is greatly affected by steam pressure and holding time. This is mainly due to the fact that glycosylated proteins covalently bind carbohydrates, which can mask the generation of fluorescence and reduce the endogenous fluorescence of tryptophan, which is consistent with the results of Corzo-Martinez et al. [34]. In addition, some proteins can be reaggregated under high-temperature action, tryptophan side chains are shielded, and high-temperature action may also oxidize tryptophan, thereby reducing fluorescence intensity. In addition, the increase in vapor pressure and the extension of the holding time can make the protein molecules more stretched, and the protein molecules covalently bind more carbohydrates, so the maximum emission wavelength of tryptophan is larger, and the fluorescence intensity is lower. Wu et al. [35] found that the fluorescence intensity of soybean protein decreased after modification with 13-hydroperoxyoctadecadienoic acid, which was thought to be caused by changes in the aggregation state and tertiary structure of proteins.

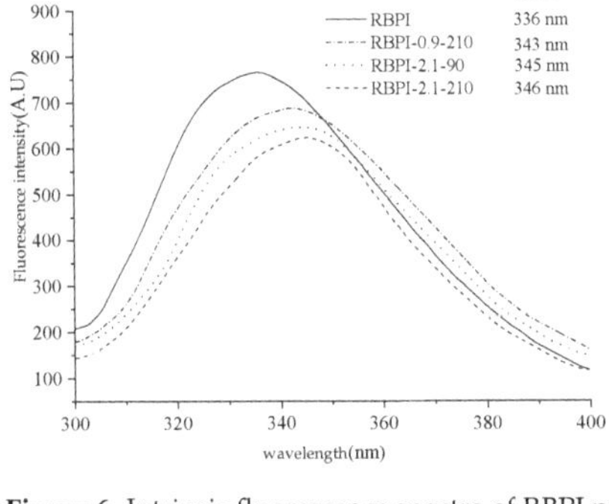

Figure 6. Intrinsic fluorescence spectra of RBPI prepared from different SFE pretreatment conditions.

The above results show that after SFE treatment, the isolated protein molecules are more stretched, the polarity of the tryptophan microenvironment increases, the protein molecules are covalently bound to the sugar chain, and the tertiary structure is changed, which is consistent with the increase in surface hydrophobicity of the isolated proteins.

2.8. Protein Particle Size Analysis

Dynamic laser scattering (DLS) was used to measure the diffusion of RBPI particles in solution as Brownian motion particles, which were converted into cumulant average particle sizes according to the Stokes-Einstein equation, as shown in Figure 7.

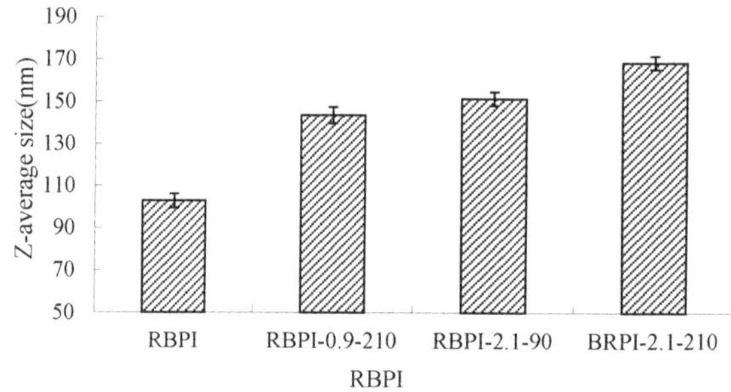

Figure 7. Average size of RBPI prepared from different SFE pretreatment conditions.

It can be seen from Figure 7 that the average particle size of RBPI from high-temperature rice bran meal was significantly affected by the Steam Flash Explosion treatment conditions and increased significantly with the increase in steam pressure and the extension of holding time ($p < 0.05$). The average particle size of the protein isolates was 102.8 nm (RBPI), and the average particle size of the protein isolates increased to 143.6 nm (RBPI-0.9-210), 151.5 nm (RBPI-2.1-90) and 168.7 nm (RBPI-2.1-210), with a maximum increase of 65.9%. This is consistent with the results shown by size exclusion chromatography, which once again proves that after SFE treatment, soluble macromolecular aggregates appear in the isolated proteins of high-temperature rice bran meal, resulting in increased protein solubility. Due to the change in the conformation of rice bran protein molecules by SFE, the existence form of spherical particles of protein molecules in an aqueous solution is destroyed, and at the same time, due to the Maillard reaction, the cross-linking between molecules is increased, resulting in an increase in the hydration radius of protein particles, so the average particle size of rice bran protein determined by dynamic laser scattering increases. Sun Tianying et al. [36] found that the isolated proteins of sunflowers underwent covalent crosslinking after heat treatment, forming macromolecular aggregates, and the protein particle size increased.

2.9. Zeta Potential Analysis

There exists an abstract boundary within the dispersion layer of the double electric layer of protein particles. When the particles move, the ions inside the boundary move with the particles, but the ions outside the boundary do not move with the particles. This boundary is called the hydrodynamic shear layer or slip surface. The potential that exists at this boundary is called the Zeta potential. Therefore, the Zeta potential of protein can be used to reflect the surface charge of protein molecules. The Zeta potential (ζ-potential) of protein at pH 7.0 was determined by a Zeta potential measuring system, the Nano ZS90 nano-particle potential analyzer, and the results are shown in Figure 8.

As can be seen from Figure 8, the Zeta potential of HTDDRB RBPI can be significantly increased by SFE, which is greatly affected by steam pressure and retention time. When the steam pressure was 2.1 MPa and the pressure holding time was 90 s, the Zeta potential of rice bran protein isolate increased by 19.4% to −24.6 mV (the Zeta of the original RBPI was −19.6 mV). When the pressure holding time was extended to 210 s under this pressure, However, the Zeta potential of RBPI decreased. This indicated that SFE could change the molecular structure of HTDDRB RBPI and increase the net charge on the molecular surface [37]. By studying the changes of endogenous fluorescence spectra of RBPI, it is concluded that the tertiary structure of RBPI is changed after sSFE, the protein molecules are more extended, and the charged polar amino acids such as tryptophan move to the surface of protein molecules, which increases the surface charge of proteins.

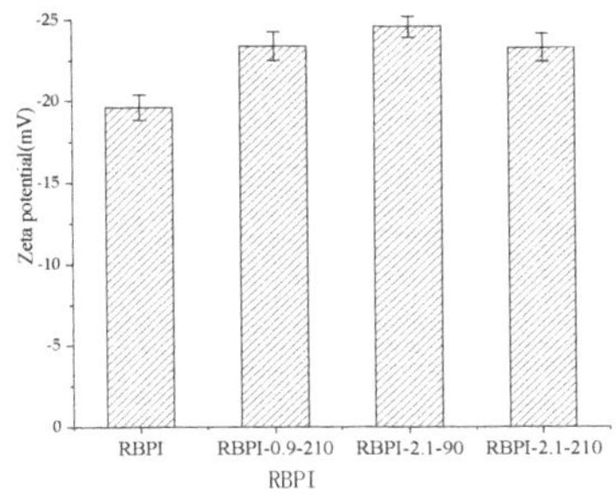

Figure 8. Zeta potential (ζ) of RBPI prepared from different SFE pretreatment conditions.

2.10. Mechanism of SFE Treatment on RBPI from HTDDRB

Because rice bran protein contains more disulfide bonds, the molecular aggregation is strong, and the heat stabilization of rice bran results in excessive denaturation of proteins in HTDDRB, and the cross-linking effect of proteins is enhanced, and the solubility of proteins is reduced, which leads to difficult extraction. The protein in HTDDRB can be subjected to various physical effects, such as high-temperature cooking, rapid movement of high-pressure steam, and instantaneous release of mechanical shear force. These effects can destroy the molecular structure of rice bran protein, reduce the α-helix content, and increase the random curling content. The secondary structure of protein changes from order to disorder, and the molecule is more extended. At the same time, mechanical shearing can depolymerize the disulfide bond covalent aggregates of protein in high-temperature rice bran meal, and stretch the peptide chains, so that the sulfols buried in the molecules are exposed. Surface hydrophobic groups, tryptophan residues, and other charged groups move to the molecular surface, and the net surface charge increases. In addition, due to high-temperature cooking and surface hydrophobic action, the oxidation of exposed sulfhydryl groups can occur so that protein molecules reaggregate, the formation of large molecules containing disulphide bond soluble aggregates, the leakage of hydrophobic groups are hidden inside the molecule so that the protein surface hydrophobicity reduced, solubility increased; At the same time, due to the extension of protein molecules, more free amino groups are exposed, which increases the probability of collision with sugar molecules, promotes the occurrence of Maillard reaction, makes protein molecules introduce more sugar chains, increases hydrophilicity, and further improves the solubility of proteins.

3. Materials and Methods

3.1. Materials

HTDDRB (130 °C, 30 min after moist heat stabilization treatment, the powder is degreased) was purchased from Heilongjiang Beidahuang Rice Industry Group Co., Ltd. (Harbin, China); 5,5′-disulfide (2-nitrobenzoic acid) (DNTB), guanidine hydrochloride, sodium dodecyl sulfate (SDS), β-mercaptoethanol (Shanghai Boao Biotechnology Co., Ltd., Shanghai, China), the above are biotechnology grade.8-aniline-1-naphthalene sulfonic acid (ANS) (Amresco Co., Ltd., Solon, OH, USA); Sodium dihydrogen phosphate, disodium hydrogen phosphate, urea, ethylenediamine tetraacetic acid (EDTA) (Tianjin Bodi Chemical Co., Ltd., Tianjin, China), all the above reagents are analytically pure. High relative molecular weight standard proteins: thyroglobulin (669 kDa), ferritin (440 kDa), aldolase (158 kDa), albumin (75 kDa), ooalbumin (44 kDa), ribonuclease (13.7 kDa), (GE-Healthcare UK Limited).

3.2. HTDDRB Steam Flash-Explosion (SFE) Treatment

The HTDDRB was crushed to 50 mesh particle size (VB-200, Beilite Vibration Machinery Co., Ltd., Xinxiang, China) by a dispersing machine (T25 basic, IKA Company, Staufen, Germany) and then divided and used. Take 400 g of crushed rice bran meal powder, according to the moisture content of 57%, and add a certain amount of distilled water to stir evenly into the SFE treatment chamber (QBS-8, Zhengdao Bioenergy Co., Ltd., Hebi, China), through a certain pressure of saturated steam and maintain pressure in a certain period of time, and then in a very short time (ms) steam flash treatment, collect steam flash samples and stored at −20 °C.

3.3. Preparation of Rice Bran Protein Isolate (RBPI)

A certain amount of HTDDRB was weighed and added with deionized water according to the solid-liquid ratio of 1:20 (g:mL), and the pH was adjusted to 9.0 by acidity meter (PB-10, Sartorius Co., Ltd., Goettingen, Germany). The meal was put into a 45 °C water bath (HWS-26, Shanghai Yiheng Scientific Instrument Co., Ltd., Shanghai, China) and extracted for 2 h under the action of mechanical agitation. The supernatant was obtained by centrifuging with a high-speed refrigerated centrifuge (TGL-20M, Hunan Xiangyi Centrifuge Instrument Co., Ltd., Changsha, China) at 4000 r/min for 15 min. The supernatant was discarded, washed, and precipitated in deionized water 3 times and centrifuged in the same way. Finally, the solution was precipitated in deionized water, the pH was adjusted to 7.0, and RBPI was obtained after freeze-drying (PDU-1100, Riken (EYELA) Devices, Tokyo, Japan). The RBPI prepared from HTDDRB treated under different SFE conditions were expressed as RBPI-0.9-150, RBPI-0.9-210, RBPI-1.5-90, RBPI-1.5-150, RBPI-2.1-90, and RBPI-2.1-210, respectively (0.9, 1.5 and 2.1 represented the pressure of saturated steam, MPa. Steam pressure; 90, 150, 210 represented pressure holding time, s), and the protein isolate prepared from untreated HTDDRB is expressed as RBPI. Mico-Kjeldahl method [38] was used to determine the protein content in rice bran protein isolate and the protein content in rice bran meal. The protein purity was calculated according to Formula (1). Table 5 shows the purity of RBPI under different SFE treatment conditions.

$$\text{purity of RBPI (\%)} = \frac{\text{Protein content in RBPI}}{\text{RBPI content}} \times 100 \tag{1}$$

Table 5. Purity of RBPI under different SFE conditions (%).

SFE Treatment Conditions	Purity of RBPI (%)
RBPI	83.6 ± 0.86
RBPI-0.9-150	79.9 ± 0.56
RBPI-0.9-210	78.1 ± 0.64
RBPI-1.5-90	77.4 ± 0.72
RBPI-1.5-150	76.6 ± 0.66
RBPI-2.1-90	75.8 ± 0.44
RBPI-2.1-210	75.0 ± 0.78

3.4. Determination of Surface Hydrophobicity (H_0)

The surface hydrophobicity (H_0) was determined by ANS (8-aniline-1-naphthalene sulfonic acid, molecular weight: 299.34) fluorescence probe [39]. The protein samples were dissolved in 10 mmol/L phosphate buffer (pH 7.0) at a protein concentration of 1.5% (*w*/*v*) (centrifugation 4000 r/min). Volumes of 10, 20, 30, 40, and 50 μL 1.5% protein solution were added to a plastic centrifuge tube containing 4 mL phosphate buffer solution, and 20 μL 8 mmol/L ANS storage solution was added before the test. The solution was shaken evenly and placed away from light for 8–15 min. The fluorescence intensity (FI) of the samples was detected by a fluorescence spectrophotometer (RF-5301PC, SHIMADZU Company, Shanghai, China). The excitation and emission wavelengths are 390 nm and

470 nm, respectively, and the slit widths are both 5 nm. The fluorescence intensity of the sample minus the reagent blank (phosphoric acid buffer) is the relative fluorescence intensity of the protein. The relative fluorescence intensity was used to plot the protein concentration, and the slope of the initial segment was used as the surface hydrophobicity index (H_0) of the protein.

3.5. Determination of Intrinsic Viscosity

The dilute solution viscosity measurement method was used. Each rice bran protein isolate was prepared into a dilute solution (0~0.008 g/mL) with deionized water as the control. A volume of 10 mL of dilute protein solution and deionized water of each concentration was placed in an Austenitic viscometer (type 1831, Shanghai Chemical Experimental Equipment Co., Ltd., Shanghai, China), and the outflow time t_0 of deionized water and the outflow time t of each dilute solution were measured in a 25 °C constant temperature water bath (HWS-26, Shanghai Yiheng Scientific Instrument Co., Ltd., Shanghai, China). The relative viscosity η_r and increased viscosity η_{sp} of each dilute solution were calculated according to Formulas (2) and (3). Time measurement accuracy is controlled within ±0.01 s, each sample is measured in parallel 4 times, and the error shall not exceed 0.1 s.

The concentration of dilute protein solution c is taken as the abscissa, and the $\ln\eta_r/c$ and η_{sp}/c are taken as the ordinate to create two straight lines, which are extrapolated to c→0, and the intercept is the intrinsic viscosity $[\eta]$.

$$\eta_r = \frac{t}{t_0} \tag{2}$$

$$\eta_{sp} = \eta_r - 1 \tag{3}$$

3.6. Differential Scanning Calorimetry (DSC) Analysis

The denaturation temperature of RBPI before and after SFE was measured by a differential scanning calorimeter (DSC-200F3, NETZSCH Instrument Company, Weimar, Germany). The temperature scanning range was 25~250 °C, and the heating rate was 5 °C/min [40].

3.7. Size Exclusion Chromatography (SEC)

The protein solution concentration was 5 mg/mL, 0.22 µm cellulose filter membrane filtration, and the loading volume was 10 uL. Agilent 1260 HPLC system and ZORBAX GF-250/50 gel column (9.4 mm × 250 mm, particle size 4 µm/6 µm) were used for the analysis (Agilent Technologies, Santa Clara, CA, USA;). The eluent was 0.05 mol/L phosphoric acid buffer solution (pH 7.0, 50 mmol/L NaCl), the flow rate was 1.0 mL/min, the column temperature was 25 °C, and the wavelength of the UV detector was 280 nm.GE high molecular weight standard protein was used to make a relative molecular weight standard curve.

3.8. Determination of Sulfhydryl and Disulfide Bonding Content

Refer to the methods of Zhang [41] and Mary et al. [42]. A mass of 100 mg rice bran protein isolate was accurately weighed and dissolved in 20 mL 0.1 mol/L pH 8.0 phosphate buffer (containing 1 mmol EDTA and 1% SDS), stirred at room temperature for 2 h, centrifuged at 10,000 r/min for 20 min (TGL-20M, Hunan Xiangyi Centrifuge Instrument Co., Ltd., Hunan, China), and the supernatant was used as the protein solution to be tested for use.

3.8.1. Free Sulfhydryl (SH_F) Determination

Take 3 mL of the protein solution to be tested and add the same volume of 0.1 mol/L pH 8.0 phosphate buffer (containing 1 mmol EDTA and 1% SDS) and 0.1 mL DNTB solution (39.6 mg DNTB dissolved in 10 mL 0.1 mol/L pH 8.0 phosphate buffer, the same below). The supernatant was mixed in a water bath at 25 °C for 1 h and centrifugation at 10,000 r/min for 20 min. The absorbance (A_F) of the supernatant was measured at 412 nm with an

Ultraviolet-visible Spectrophotometer (T6, Beijing Puchan Universal Instrument Co., Ltd., Beijing, China). The content of free sulfhydryl (SHF) was calculated according to Formula (4).

3.8.2. Total Sulfhydryl Determination (SH_T)

Take 1 mL protein solution to be tested, add 0.05 mL β-mercaptoethanol and 4 mL urea-guanidine hydrochloride solution (0.1 mol/L pH 8.0 phosphate buffer, including 8 mol/L urea and 5 mol/L guanidine hydrochloride), and then add 10 mL 12% TCA solution in 25 °C water bath for 1 h. After a water bath at 25 °C for 1 h, centrifuge at 5000 r/min for 10 min (TGL-20M, Hunan Xiangyi Centrifuge Instrument Co., Ltd., Changsha, China), discard the clear liquid, wash precipitate with 12% TCA to remove β-mercaptoethanol, and repeat twice. A volume of 10 mL of 0.1 mol/L pH 8.0 phosphate buffer (containing 1 mmol EDTA and 1% SDS) and 0.08 mL DNTB solution were added to the precipitation. After the solution was fully dissolved, the water bath at 25 °C for 1 h was centrifuged at 10,000 r/min for 20 min. The absorbance (A_T) of the supernatant was measured at 412 nm with an Ultraviolet-visible Spectrophotometer (T6, Beijing Puchan Universal Instrument Co., Ltd., Beijing, China). The reaction solution without DNTB was used as the control. The total sulfhydryl (SH_T) content and disulfide (SS) content were calculated according to Formula (5) and (6), respectively.

$$SH_F(\mu\text{mol/g}) = 73.53 \times A_F \times \frac{D}{C} \tag{4}$$

$$SH_T(\mu\text{mol/g}) = 73.53 \times A_T \times \frac{D}{C} \tag{5}$$

$$SS(\mu\text{mol/g}) = \frac{SH_T - SH_F}{2} \tag{6}$$

Formula: C—protein concentration, mg/mL; D—dilution factor.

3.9. Circular Dichroic (CD) Spectrum

According to the method of Shahabadi et al. [43], a certain amount of RBPI was accurately measured by freeze-dried samples and dissolved in 10 mmol/L pH 7.0 phosphate buffer to prepare 0.2 mg/mL protein solution. The solution was measured on a circular dichroic spectrometer (J-810, Japan Spectroscopic JASCO Corporation, Tokyo, Japan). Determination conditions: The optical diameter was 2 mm, the temperature was 25 °C, the sensitivity was 20 mdeg/cm, the scanning speed was 100 nm/min, the width was 1 nm, the scanning range was 190~255 nm, the response time was 0.5 s, the scanning was repeated 5 times to take the mean value, the pH 7.0 phosphate buffer was used as blank. The Spectra Manager driver software workstation of the machine provides the secondary structure of the protein.

3.10. Determination of Endogenous Fluorescence of RBPI

RBPI, RBPI-0.9-210, RBPI-2.1-90, and RBPI-2.1-210 were used as research objects to determine the endogenous fluorescence spectra of tryptophan by a fluorescence spectrophotometer (RF-5301PC, SHIMADZH Company of Japan, Kyoto, Japan). Each protein isolate was dissolved in a phosphate buffer solution (pH 7.0, 0.01 mol/L) and prepared into 0.15 mg/mL solution. The excitation wavelength is 295 nm, the scanning range of the emission spectrum is 300~400 nm, the slit width of excitation and emission is 5 nm, and the scanning speed is 10 nm/s.

3.11. Dynamic Laser Light Scattering (DLS) Determines Protein Particle Size

A certain amount of RBPI was accurately measured from freeze-dried samples, dissolved in 10 mmol/L pH 7.0 phosphate buffer, and filtered by 0.22 μm cellulose filter membrane to prepare 1 mg/mL protein solution. The diffusion of Brownian motion of protein particles was measured by the DLS component of a Malvern Nano ZS90 nanoparticle potential analyzer, which was converted into cumulative Z-average Size according to the Stokes-Einstein equation using DTS software [44]. The distribution characteristics of dynamic light scattering data at a 90° scattering Angle were analyzed by the cumulant analysis method.

3.12. Determination of Zeta Potential

Prepare 1 mg/mL protein solution with the same method as 3.11. The Zeta potential (Zeta—potential) of protein was determined by the Zeta potential measuring system of a Malvern Nano ZS90 nano size potential analyzer (Malvern Ltd., Royal Leamington Spa, UK). Electrophoretic mobility (UE) was measured using a combination of electrophoresis and laser Doppler velocimetry (laser Doppler electrophoresis), and Zeta potential was calculated based on the Henry equation.

3.13. Statistical Analysis

All experimental data were averaged over three replicates, and SPSS 24.0 data analysis software was used for one-way ANOVA and significance analysis, and experimental data processing and plotting were performed by Excel 2010 and OriginPro 9.0 software.

4. Conclusions

HTDDRB was treated by SFE, the surface hydrophobicity and thermal stability of the isolated proteins decreased significantly, the characteristic viscosity increased significantly, the α-helix content decreased, but the random curling content increased, and the secondary structure of the proteins tended to be disordered. The analysis of relative molecular mass distribution, endogenous fluorescence spectrum, free sulfhydryl and disulfide bonds, particle size, and the Zeta potential of binding proteins showed that after steam flash treatment, the disulfide bond covalent aggregates of rice bran protein isolates were depolymerized, the molecules extended, the sulfhydryl groups buried in the molecules were exposed, and the hydrophobic surface groups and tryptophan residues moved to the molecular surface. The net surface charge increases; At the same time, due to high-temperature cooking and surface hydrophobic action, some proteins aggregate, forming soluble macromolecular aggregates, hydrophobic groups leaked inside the molecules, and surface hydrophobicity decreased while protein solubility increased.

Author Contributions: Data curation, Z.N.; investigation, H.B.; writing-review and editing Y.W.; writing-review and editing, Y.G.; conceptualization and supervision, Y.M. All authors have read and agreed to the published version of the manuscript.

Funding: This research was supported by Natural Science Foundation of Heilongjiang Province of China (Grant No. LH2020C078).

Institutional Review Board Statement: Not applicable.

Informed Consent Statement: Not applicable.

Data Availability Statement: Not applicable.

Conflicts of Interest: The authors declare no conflict of interest.

References

1. Li, X.M.; Hu, Z.X.; Zhang, W.N.; Qi, Y.T.; Zhang, Y.P.; Du, Y.; Zuo, C. Characterization and application of carbon dots from rice bran protein. *Chinese Grease* **2022**, *47*, 58–64.
2. Chen, J.S.; Liu, C.C.; Shi, B.; Chai, Y.; Han, N.; Zhu, M.; Bian, H. Overexpression of HvHGGT Enhances Tocotrienol Levels and Antioxidant Activity in Barley. *J. Agric. Food. Chem.* **2017**, *65*, 5181–5187. [CrossRef]
3. Li, X.F.; Shen, D.; Huang, J.K.; Lu, Y.J. Research progress on functional active components and its efficacy in rice bran. *Food Industry Technology* **2022**, *43*, 466–474.
4. Lian, Z.; Liu, X.L. Extraction and functional properties of four proteins from reeling silkworm pupa. *Food Sci. Technol.* **2018**, *43*, 280–285.
5. Zhou, X.T.; Liu, L.; Wang, L.; Liu, T.; Wu, X. Proteomic study of Chinese black-bone silky fowl and the ring-necked pheasant egg white by iTRAQ technique. *LWT* **2021**, *150*, 111936. [CrossRef]

6. Xiao, D.Z.; Chong, H.Y.; Yu, X.W.; Mei, L.S.; Guo, P.Y. Effect of limited enzymatic hydrolysis on the structure and emulsifying properties of rice bran protein. *J. Cereal Sci.* **2019**, *85*, 168–174.
7. Li, G.T.; Fan, Z. Amylopectin molecular structure in relation to physicochemical properties of quinoa starch. *Carbohyd. Polym.* **2017**, *164*, 396–402. [CrossRef] [PubMed]
8. Cheng, Y.H.; Mu, D.C.; Feng, Y.Y.; Xu, Z.; Wen, L.; Chen, M.L.; Ye, J. Glycosylation of rice protein with dextran via the Maillard reaction in a macromolecular crowding condition to improve solubility. *J. Cereal Sci.* **2022**, *103*, 103374. [CrossRef]
9. Wang, L.; Pang, T.R.; Kong, F.; Chen, H.Z. Steam Explosion Pretreatment for Improving Wheat Bran Extrusion Capacity. *Foods* **2022**, *11*, 2850. [CrossRef] [PubMed]
10. Wan, F.C.; Hou, C.Y.; Luo, K.Y.; Cheng, A.W. Steam explosion enhances phenolic profiles and antioxidant activity in mung beans. *Food Sci. Nutr.* **2022**, *10*, 1039–1050. [CrossRef]
11. Orevi, T.; Sørensen, S.J.; Kashtan, N. Droplet size and surface hydrophobicity enhance bacterial plasmid transfer rates in microscopic surface wetness. *ISME Comm.* **2022**, *2*, 72. [CrossRef]
12. Dela Cruz-Torres, L.F.; Rodríguez-Celestino, V.; Centeno-Leija, S.; Serrano-Posada, H.; Ceballos-Mangana, S.G.; Aguilar-Padilla, J.; Mancilla-Margalli, N.A.; Osuna-Castro, J.A. Development of a rapid, high-sensitivity, low-cost fluorescence method for protein surface hydrophobicity determination using a Nanodrop fluorospectrometer. *Food Chem.* **2022**, *396*, 133681. [CrossRef]
13. Sun, C.L.; Du, W.; Wang, B.Q.; Dong, B.; Wang, B.G. Research progress of near-infrared fluorescence probes based on indole heptamethine cyanine dyes in vivo and in vitro. *BMC Chem.* **2020**, *14*, 21. [CrossRef] [PubMed]
14. Zhang, Y.P.; Yang, R.J.; Zhao, W.; Hua, X.; Zhang, W.B. Application of high density steam flash-explosion in protein extraction of soybean meal. *J. Food Eng.* **2013**, *116*, 430–435. [CrossRef]
15. Nayak, A.; Genot, C.; Meynier, A.; Dorlando, A.; Capron, I. Impact of process and physico-chemical conditions on the formation of curcumin-whey protein composite particles capable to stabilize food-compatible oil in water emulsions. *LWT* **2022**, *153*, 112421. [CrossRef]
16. Na, Z.G.; Wang, H.M.; Yan, B.J.; Zhang, L.; Ma, Y.Q. Effect of Steam Flash Explosion on Maillard Reaction of High-Temperature Denatured Defatted Rice Bran. *Food. Industry* **2020**, *41*, 76–82.
17. Achour, A.; Boye, J.I.; Yaylayan, V.A.; Yeboah, F.K. Functional Properties of Glycated Soy 11S Glycinin. *J. Food Sci.* **2005**, *70*, 269–274.
18. Gasymov, O.K.; Glasgow, B.J. ANS fluorescence: Potential to augment the identification of the external binding: Sites of proteins. *Acta Biochim. Biophys. Sin.* **2007**, *1774*, 403–411. [CrossRef] [PubMed]
19. Wang, C.F.; Hao, Y.; Wang, Y.; Song, H.J.; Hussain, S.; Gao, R.X.; Gao, L.Y.; He, Y.L.; Zheng, G.D.; Tang, Y.H. Multiwall Carbon Nanotubes Non-covalently Functionalized by Porphyrin-Sn Networks for Protein Adsorption. *ACS Appl. Nano Mater.* **2021**, *4*, 2345–2350. [CrossRef]
20. Jia, Y.J.; Yan, X.Y.; Li, X.T.; Zhang, S.; Huang, Y.Y.; Zhang, D.M.; Li, Y.; Qi, B.K. Soy protein–phlorizin conjugate prepared by tyrosinase catalysis: Identification of covalent binding sites and alterations in protein structure and functionality. *Food Chem.* **2023**, *404*, 134610. [CrossRef]
21. Kinsella, J.E. Functional properties of proteins: Possible relationships between structure and function infoams. *Food Chem.* **1981**, *7*, 273–288. [CrossRef]
22. Zhang, Y.P.; Yang, R.J.; Zhang, W.N.; Hu, Z.X.; Zhao, W. Structural characterization and physicochemical properties of protein extracted from soybean meal assisted by steam flash-explosion with dilute acid soaking. *Food Chem.* **2017**, *219*, 48–53. [CrossRef]
23. Mir, N.A.; Riar, C.S.; Singh, S. Improvement in the functional properties of quinoa (Chenopodium quinoa) protein isolates after the application of controlled heat-treatment: Effect on structural properties. *Food Struct.* **2021**, *28*, 100189. [CrossRef]
24. Gao, Y.X.; Sun, Y.; Zhang, Y.; Sun, Y.K.; Jin, T. Extrusion Modification: Effect of Extrusion on the Functional Properties and Structure of Rice Protein. *Processes* **2022**, *10*, 1871. [CrossRef]
25. Wu, W.; Cai, Y.J.; Wu, X.J.; Meng, Q.; Ye, J.F. Effect of Rice Bran Storage Time on the Structure of Rice Bran Protein. *Mod. Food Technol.* **2017**, *33*, 1–9.
26. Xia, N.; Wang, J.; Yang, X.; Yin, S.; Qi, J.; Hu, L.; Zhou, X. Preparation and characterization of protein from heat-stabilized rice bran using hydrothermal cooking combined with amylase pretreatment. *J. Food Eng.* **2012**, *110*, 95–101. [CrossRef]
27. David, M.R.; Sarah, B.J.; Naomi, T.D.; Francois, A.; Matthieu, R.; Jonathan, D. Electronic Circular Dichroism Spectroscopy of Proteins. *Chem* **2019**, *5*, 2751–2774.
28. Sze, K.W.C.; Kshirsagar, H.H.; Venkatachalam, M.; Sathe, S.K. A circular dichroism and fluorescence spectrometric assessment of effects of selected chemical denaturants on soybean (*Glycine max* L.) storage proteins glycinin (11S)and β-conglycinin(7S). *J. Agric. Food Chem.* **2007**, *55*, 8745–8753.
29. Damodaran, S. *Food Proteins and Their Applications*; Taylor and Francis: Abingdon, UK; CRC Press: Boca Raton, FL, USA, 2017.
30. Yagi, U.M.; Kato, K. Conformational Variability of Amyloid-β and the Morphological Diversity of Its Aggregates. *Molecules* **2022**, *27*, 4787. [CrossRef]
31. Shutova, T.; Deikus, T.; Irrgang, K.D.; Klimov, V.V.; Renger, G. Origin and properties of fluorescence emission from the extrinsic 33kDa manganese stabilizing protein of higher plant water oxidizing complex. *BBA-Bioenergetics* **2001**, *1504*, 371–378. [CrossRef]
32. Shi, X.L.; Fang, Y.Y.; Wei, Q.N.; Wang, J.K.; Si, Y.X.; Chen, Z.F.; Qian, G.Y.; Yin, S.J.; Wang, Q. A heat-resistant β-mannanase derived from the Thermothelomyces thermophilus. *Acta Microbiol. Sin.* **2021**, *61*, 2815–2828.

33. Jia, X.; Li, L.C.; Teng, J.W.; Li, M.J.; Long, H.; Xia, N. Glycation of rice protein and d-xylose pretreated through hydrothermal cooking-assisted high hydrostatic pressure: Focus on the structural and functional properties. *LWT* **2022**, *160*, 113194. [CrossRef]
34. Corzo-Martínez, M.; Moreno, F.J.; Olano, A.; Villamiel, M. Structural characterization of bovine beta-lactoglobulin-galactose/tagatose Maillard complexes by electrophoretic, chromatographic, and spectroscopic methods. *J. Agric. Food Chem.* **2008**, *56*, 4244–4252. [CrossRef]
35. Wu, W.; Hua, Y. Structural modification of soy protein by the lipid peroxidation product acrolein. *LWT-Food Sci. Technol.* **2010**, *43*, 133–140. [CrossRef]
36. Sun, T.Y.; Cheng, H.; Chen, K.Y.; Ren, J. Effects of Heat on the Structural Properties of Sunflower Protein Isolate. *Chin. Condiments* **2021**, *46*, 69–71+76.
37. Feng, H.Y.; Jin, H.; Gao, Y.; Yan, S.Q.; Zhang, Y.; Zhao, Q.S.; Xu, J. Effects of freeze-thaw cycles on the structure and emulsifying properties of peanut protein isolates. *Food Chem.* **2020**, *330*, 127215. [CrossRef] [PubMed]
38. Rizvi, N.B.; Aleem, S.; Khan, M.R.; Ashraf, S.; Busquets, R. Quantitative Estimation of Protein in Sprouts of Vigna radiate (Mung Beans), Lens culinaris (Lentils), and Cicer arietinum (Chickpeas) by Kjeldahl and Lowry Methods. *Molecules* **2022**, *27*, 814. [CrossRef]
39. Wang, J.Y.; Yang, Y.L.; Tang, X.Z.; Ni, W.X.; Zhou, L. Effects of pulsed ultrasound on rheological and structural properties of chicken myofibrillar protein. *Ultrason. Sonochem.* **2017**, *38*, 225–233. [CrossRef]
40. Sun, H.Y.; Deng, G.; Zhou, Y.; Wei, Z.W. Recent progress in protein and protein-related systems using DSC techniques. *Chin. Sci. Bull.* **2016**, *61*, 3091–3099. [CrossRef]
41. Zhang, Y.Y.; Li, Y.L.; Li, S.Y.; Zhang, H.; Ma, H.L. In Situ Monitoring of the Effect of Ultrasound on the Sulfhydryl Groups and Disulfide Bonds of Wheat Gluten. *Molecules* **2018**, *23*, 1376. [CrossRef]
42. Mary, A.O.; Takafumi, I.; Roberta, N.; Nobuyuki, M.; Shigeru, U.; Evelyn, M. Introduction of sulfhydryl groups and disulfide linkage to mungbean 8Sα globulin and effects on physicochemical and functional properties. *Food Res. Int.* **2012**, *45*, 277–282.
43. Shahabadi, N.; Ghaffari, L.; Mardani, Z.; Shiri, F. Multi-Spectroscopic and Theoretical Analyses of Human Serum Albumin Binding to a Water-Soluble Zinc(II) Complex including β-Amino Alcohol. *Chem. Select* **2022**, *7*, e202200645. [CrossRef]
44. Larichev, Y.V. Extended Possibilities of Analysis for Supported Metal Catalysts and Nanocomposites by Dynamic Light Scattering. *Kinet. Catal.* **2022**, *63*, 299–605. [CrossRef]

molecules

Article

Effects of Different Denaturants on the Properties of a Hot-Pressed Peanut Meal-Based Adhesive

Yang Qu †, Qin Guo †, Tian Li, Hongzhi Liu * and Qiang Wang *

Institute of Food Science and Technology, Chinese Academy of Agricultural Sciences, Key Laboratory of Agro-Products Processing, Ministry of Agriculture, Beijing 100194, China; 13999411257@163.com (Y.Q.); guoqin@caas.cn (Q.G.); 18811777506@163.com (T.L.)
* Correspondence: lhz0416@126.com (H.L.); wangqiang06@caas.cn (Q.W.); Tel./Fax: +86-10-62815837 (Q.W.)
† These authors contributed equally as first authors.

Abstract: Plant protein-based adhesives could fundamentally solve the problem of formaldehyde-based adhesive releasing formaldehyde, but enhancing bonding strength and water resistance is a necessary measure to realize practical applications. In this study, the effects of different denaturants on the properties of a hot-pressed peanut meal (HPM)-based adhesive before and after crosslinking were studied. Papain, sodium dodecyl sulfate (SDS), urea and crosslinker-polyamide epichlorohydrin (PAE) were used to prepare HPM-based adhesives. The functional groups, bonding strength, thermal behaviors, mass loss, moisture uptake value, viscosity and fracture surface of adhesive samples were analyzed. As a result, (1) papain was used to break HPM protein (HPMP) into polypeptide chains and to reduce the water resistance. (2) SDS and urea unfold the HPMP molecule and expose internal hydrophobic groups to improve the water resistance of the adhesive. (3) A denser network structure was formed by PAE and HPMP molecules, which significantly improved the bonding strength and water resistance of adhesives. In particular, after SDS denaturation and PAE crosslinking, compared with pure HPM adhesive, the wet shear strength increased by 96.4%, the mass loss and moisture uptake value reduced by 41.4% and 69.4%, and viscosity increased by 30.4%. This work provided an essential guide to design and prepare HPM-based adhesives.

Keywords: hot-pressed peanut meal; denaturation; network structure; crosslinking; water resistance

Citation: Qu, Y.; Guo, Q.; Li, T.; Liu, H.; Wang, Q. Effects of Different Denaturants on the Properties of a Hot-Pressed Peanut Meal-Based Adhesive. *Molecules* **2022**, *27*, 4878. https://doi.org/10.3390/molecules27154878

Academic Editors: Shudong He, Wei Xu and Muhammad H. Aludatt

Received: 30 June 2022
Accepted: 29 July 2022
Published: 30 July 2022

1. Introduction

Formaldehyde-based adhesives are widely used in the manufacturing of plywood, particleboard and fiberboard, and they are derived from non-renewable fossil sources [1]. In addition, formaldehyde-based adhesive release formaldehyde and free phenol in the process of preparation, transportation and application, which can harm the human body [2]. Therefore, plant proteins [3], lignin [4], starch [5] and other sustainable bio-based raw materials have been used as alternatives to develop new wood adhesives and have a high potential for use in industry. Most studies have focused on the use of different chemical methods to enhance the mechanical performance and water resistance of soybean meal adhesives in recent years [6]. However, soybean meal is mainly used for feed and is affected to the soybean market. Therefore, it is important to make full use of existing protein resources that are considered waste to alleviate protein shortages and to promote sustainable development.

Hot-pressed peanut meal (HPM) is a by-product of pressing peanut oil at high temperature from peanut kernels. [7]. It cannot be used in feed or food processing industry due to the content of aflatoxin exceeding the standard, and thus can only be used as waste [8]. HPM contains more than 45% protein, 87% of which is globulin, which is mainly composed of arachin (glycinin) and conarachin (vicilin) [9]. According to previous reports, HPM protein (HPMP) and soybean protein are similar in their amino acid composition and protein structure; they can both react with compounds to produce adhesives [10].

However, HPM adhesive cannot be used in wet environments due to molecular physical entanglement and hydrogen bonding [11]. In order to realize the practical application of HPM adhesives, the structure and properties of natural proteins must be modified to improve their reactivity [12]. Common strategies used in previous reports include denaturation [13], crosslinking [14], and nano-material modification [15]. However, there have been no reports on the effects of different denaturation methods on the properties of HPM adhesive. Therefore, this study can help to develop a more efficient preparation method for HPM-based adhesives.

In this study, different denaturants (papain, SDS, urea) were used to improve the properties of HPM-based adhesives. After the papain treatment, more hydrophilic groups were exposed because HPMP was broken into peptides, which increased the reaction sites and facilitated the cross-linking reaction between HPM and the cross-linking agent [16]. SDS could be inserted into HPMP molecules to break hydrogen bonds and expose hydrophobic groups [17]. The intramolecular hydrogen bonds were destroyed by urea, promoting the unfolding of HPMP molecules [18]. Based on this consideration, three different HPM-based adhesives were prepared through a reaction with PAE. Furthermore, the effects of different denaturation methods on the functional groups, bonding strength, thermal behaviors, mass loss (ML), moisture uptake value (MUV), viscosity and fracture surface of different HPM-based adhesives were studied by means of FTIR, DTG, bonding strength evaluation and viscosity analysis. These results provide a theoretical basis for the further development of HPM-based adhesives.

2. Materials and Methods

2.1. Materials

HPM (200 mesh, 48.2% protein, 25.8% carbohydrates, 6.39% moisture content, 6.96% ash, 6.87% fiber and 5.14% fat) was obtained from Shandong LuHua grain, oil and Food Co., Ltd. (Shandong, China). SDS, urea and papain were obtained from Shanghai Macklin Biochemical Co., Ltd. (Shanghai, China). Polyamide epichlorohydrin resin (PAE, 25 wt%) was obtained from Zhejiang ChuanHua Huayang Chemical Co., Ltd. (Zhejiang, China). Poplar veneer (25 cm × 25 cm × 0.17 cm, 9.4% moisture content) was obtained from Wenan, Hebei province, China.

2.2. Preparation of HPM-Based Adhesive

HPM powder (20 g) and deionized water (55 g) were stirred at 25 °C for 20 min to prepare pure HPM adhesive. Then, 0.8 g of the denaturant agent (papain, SDS and urea) was added into pure HPM adhesive and continuously stirred for 20 min, and the mixtures were marked as HPM-papain, HPM-SDS and HPM-urea adhesive, respectively. Then, PAE (16g) was mixed with different adhesives, and stirred for 10 min at 25 °C, and the mixtures were marked as HPM-papain-PAE, HPM-SDS-PAE and HPM-urea-PAE adhesives, respectively. The formulations are shown in Table 1.

Table 1. Formulations used for adhesive samples.

Adhesive Sample	HPM (g)	Distilled Water (g)	Denaturant (g)	PAE (g)
Pure HPM	20	55	0	0
HPM-Papain	20	55	0.8	0
HPM-SDS	20	55	0.8	0
HPM-Urea	20	55	0.8	0
HPM-PAE	20	55	0	16
HPM-Papain-PAE	20	55	0.8	16
HPM-SDS-PAE	20	55	0.8	16
HPM-Urea-PAE	20	55	0.8	16

2.3. Fourier Transform Infrared (FTIR) Spectroscopy

The samples of adhesive were freeze-dried for 72 h and ground to 200 mesh. Before the test, the sample powder was mixed with KBr at 1:100 and pressed into thin slices. The FTIR spectra were obtained at a wave number from 400 to 4000 cm^{-1} during 64 scans, with 2 cm^{-1} resolution.

2.4. Three-Ply Plywood Preparation and Bonding Strength Test

Three layers of plywood were prepared with different adhesive, and the amount of adhesive applied on each layer was 220 g/m^2. The veneer coated with adhesive was placed in the middle as the core layer, and the grain was perpendicular to the upper and lower sides of the veneer. It was stored at room temperature for 5 min after assembly. Hot pressing was performed at 120 °C, 1.2 MPa and 100 s/mm. The prepared plywood was tested after being left at room temperature for 24 h. A total of 30 specimens (adhesive joint area of 25 mm × 25 mm) were cut from five different plywood samples to determine the dry shear strength (DS) and wet shear strength (WS) according to GB/T 9846-2015 [19]. The DS was tested using a universal testing machine with a crosshead speed of 5 mm/min. For WS, the specimen was immersed in water at 63 °C for 3 h and cooled to room temperature for 10 min, then the WS was measured.

2.5. Thermogravimetry (TGA)

Freeze-dried samples (8 mg dry weight) were weighed in an aluminum crucible. The temperature was increased from 30 °C to 500 °C under a nitrogen flow of 10 mL/min, the heating rate was 20 °C/min [20].

2.6. Mass Loss (ML) and Moisture Uptake Value (MUV) Test

ML and MUV measurement were performed according to the scheme of Qu et al. [21]. The adhesive samples were oven-dried at 110 °C to a constant weight (M_a). The cured adhesive samples were immersed in water at room temperature. After 48 h, the adhesive samples were taken out and dried to a constant weight (M_b). The formulas were as listed below:

$$ML\ (\%) = [(M_b - M_a)/M_b] \times 100\% \quad (1)$$

To determine the dry mass, the adhesive samples (8 g) were desiccated to a constant weight at 105 °C. Next, we placed the adhesive samples in a constant-temperature and -humidity incubator at 50 °C and 85% humidity (saturated KCl solution). We recorded the mass of the adhesive sample every 2 h and continued to measure until the sample reached a constant weight. M_1 and M_2 were the mass of the samples after MUV and after drying, respectively. The equation was rendered thus:

$$\text{Moisture uptake}\ (\%) = [(M_1 - M_2)/M_2] \times 100\% \quad (2)$$

2.7. Viscosity Test

The viscosity of adhesive samples was measured at room temperature using a viscometer (DV-III, Ultra, Middleboro, MA, USA), and each sample was measured three times.

2.8. Scanning Electron Microscopy (SEM)

The cured adhesive sample was adhered to the metal table, then the sample was sprayed with gold using JFC-1100E ion sputter (JEOL, Tokyo, Japan). The micromorphology of the adhesive fracture surface was observed using a field-emission scanning electron microscope (Hitachi SU8010).

3. Results and Discussion

3.1. FTIR Analysis

The structural changes of different adhesive samples were studied by FTIR (Figure 1). The characteristic absorption peaks of HPM were observed at 1654, 1542 and 1342 cm^{-1}, which belong to amide C=O stretching (amide I), N-H bending (amide II), C–N stretching and N–H vibrations (amide III), respectively [9]. No new characteristic peaks appeared in the infrared spectrum after adding papain (HPM-Papain adhesive), SDS (HPM-SDS adhesive) and urea (HPM-urea adhesive), demonstrating that the primary structure of the protein molecule had not been changed [22]. Incorporating SDS into the adhesive led to a blue shift of the amide I, amide II, and amide III peaks from 1654, 1542, and 1242 cm^{-1} to 1664, 1545, and 1244 cm^{-1} for the HPM-SDS adhesive, respectively. Similar phenomena were observed in the HPM-urea adhesive, indicating that the intermolecular hydrogen bond of protein was destroyed and more active groups were exposed [16]. With the addition of PAE, a significant blue shift occurred in the characteristic peak of 3369 cm^{-1} (O–H and N–H bending vibrations), indicating that the original hydrogen bond was destroyed and physical bonding was reconstructed [23]. The peak area (COO–) decreased at 1391 cm^{-1}, and a new characteristic peak of the carbonyl group appeared at 1742 cm^{-1}, which was attributed to esterification of the azacyclobutane group with the carbonyl group of HPMP [21]. In addition, the activation energy of the amino group was lower than that of the carbonyl reaction, which means that it forms a network structure with PAE. The network structure helped to decrease the number of hydrophilic groups in the adhesive and improve the water resistance. The mechanism of the cross-linking is presented in Figure 2.

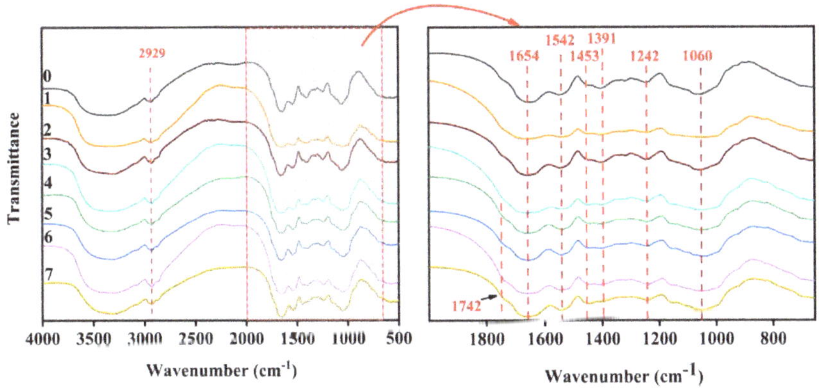

Figure 1. Attenuated total reflection FTIR spectroscopic results from the cured adhesive samples: (0) HPM adhesive, (1) HPM-papain adhesive, (2) HPM-sodium dodecyl sulfate (SDS) adhesive, (3) HPM-urea adhesive, (4) HPM-PAE adhesive, (5) HPM-papain-PAE adhesive, (6) HPM-SDS-PAE adhesive, (7) HPM-urea PAE adhesive.

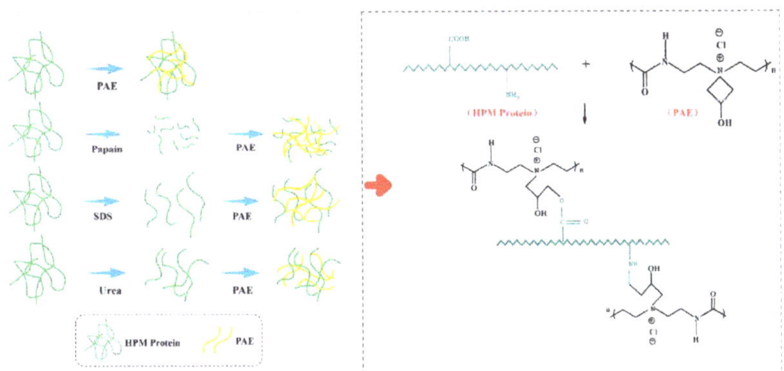

Figure 2. The denaturation and crosslinking mechanism of HPMP molecule.

3.2. Bonding Strength Measurement

The DS and WS of the different adhesive samples are presented in Figure 3. The HPM adhesive had a DS of 0.97 MPa and a WS of 0.57 MPa. After adding papain, the HPMP was broken down into polypeptide, the DS and WS of plywood with HPM-papain adhesive were reduced by 10.3% (0.87 MPa) and 40.4% (0.34 MPa) [24]. The ordered structure of native HPMP was denatured as SDS concentration increased, so the DS and WS of HPM-SDS adhesive were significantly increased by 61.9% (1.57 MPa) and 24.6% (0.74 MPa). When the urea was added, the DS (1.49 MPa) and WS (0.67 MPa) of HPM-urea adhesive was also increased significantly. This was due to the secondary structure of the HPMP unfolding, which was conducive to the exposure of hydrophobic groups. The WS of HPM-papain-PAE adhesive, HPM-SDS-PAE adhesive and HPM-urea-PAE adhesive was increased by 57.9% (0.9 MPa), 96.4% (1.12 MPa) and 93% (1.1 MPa) with the addition of PAE, which met the plywood (type II) of Chinese National Standard (GB/T 9846-2015). This was attributed to PAE being distributed into HPM and forming a network structure with the active group (–NH_2, –COOH) [25]. This result was confirmed by comparison with the HPM-PAE adhesive.

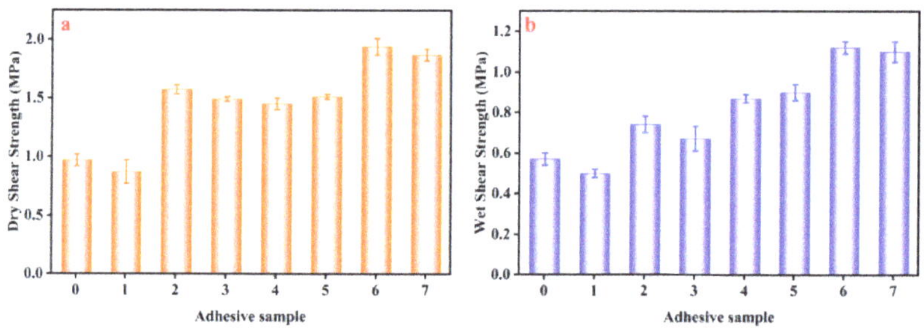

Figure 3. The (**a**) dry shear strength and (**b**) wet shear strength of different adhesive samples: (0) HPM adhesive, (1) HPM-papain adhesive, (2) HPM-sodium dodecyl sulfate (SDS) adhesive, (3) HPM-urea adhesive, (4) HPM-PAE adhesive, (5) HPM-papain-PAE adhesive, (6) HPM-SDS-PAE adhesive, (7) HPM-urea-PAE adhesive.

3.3. TGA Analysis

The thermal behaviors of different adhesive samples are shown in Figure 4. The three stages of thermal degradation of the adhesive could be clearly observed (Figure 4a). The first stage occurred over the 123–222 °C temperature range, while the weight loss was attributable to the evaporation of residual moisture [26]. The second stage (222–282 °C) had the most obvious mass loss, which was the decomposition of small molecular substances [27]. Finally, the third stage (282–362 °C) corresponded to adhesive skeleton degradation, including the degradation of peptide bonds and the cleavage of C–N, C–O [28]. After papain treatment, the degradation peak of adhesive 1 moved to a low temperature (299.45 °C), which indicated that the thermal stability decreased. Under SDS and urea, the thermal decomposition temperature of adhesive 2 and adhesive 3 increased by 3.5% (316.69 °C) and 2.8% (314.53 °C), respectively, the improvement could be attributed to the physical enhancement [29]. The thermal degradation peaks of adhesive 5 (324.89 °C), adhesive 6 (325.78 °C) and adhesive 7 (325.99 °C) moved toward higher temperatures with the addition of PAE. Combined with the FTIR analysis, the network structure formed by PAE and HPM could enhance the thermal stability of the adhesive.

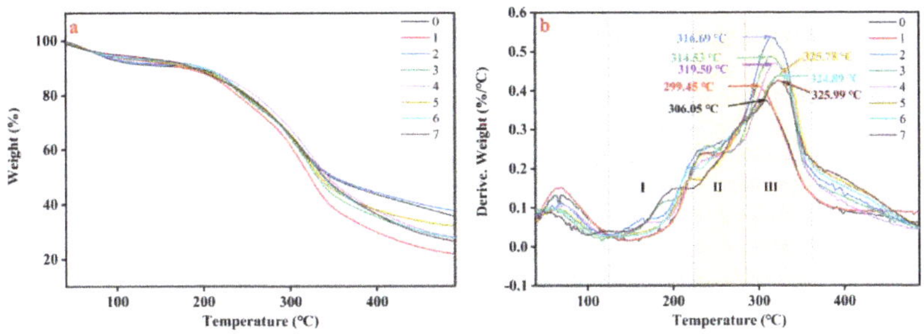

Figure 4. (**a**) Derivative thermogravimetric (DTG) and (**b**) thermogravimetric and curves of the different adhesive samples: (0) HPM adhesive, (1) HPM-papain adhesive, (2) HPM-sodium dodecyl sulfate (SDS) adhesive, (3) HPM-urea adhesive, (4) HPM-PAE adhesive, (5) HPM-papain-PAE adhesive, (6) HPM-SDS-PAE adhesive, (7) HPM-urea-PAE adhesive.

3.4. ML and MUV Measurement

The ML and MUV were significantly and negatively correlated with the adhesive's crosslinking degree [30]. Figure 5 showed the ML and MUV of HPM adhesive were 54.37% and 9.25%, respectively. In addition, papain exposed the active hydrophilic groups inside the HPMP molecule, so that the ML and MUV of HPM-papain were increased to 73.25% (Figure 5a) and 12.8% (Figure 5b). As expected, the ML and MUV of HPM-SDS adhesive and HPM-urea adhesive decreased slightly. The ML of HPM-PAE adhesive, HPM-papain-PAE adhesive, HPM-SDS-PAE adhesive, and HPM-urea-PAE adhesive decreased to 44.37%, 38.67%, 31.68% and 34.44%, respectively. The MUV demonstrated the same tendency. This showed that the network structure formed by PAE and HPM significantly improved the water resistance of the adhesive. These results are consistent with the bonding strength analysis.

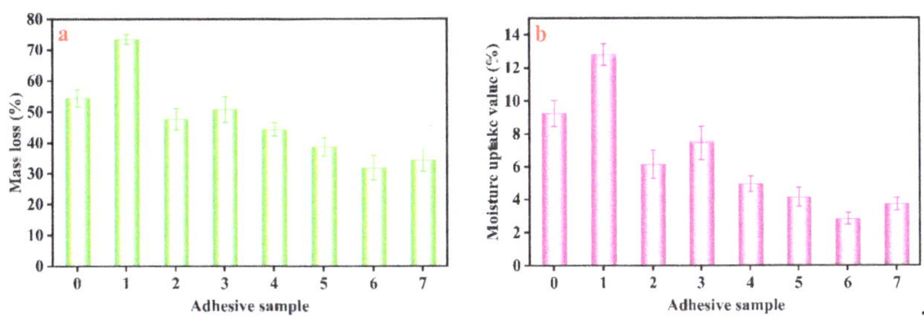

Figure 5. The (**a**) mass loss and (**b**) moisture uptake value of the different adhesive samples: (0) HPM adhesive, (1) HPM-papain adhesive, (2) HPM-sodium dodecyl sulfate (SDS) adhesive, (3) HPM-urea adhesive, (4) HPM-PAE adhesive, (5) HPM-papain-PAE adhesive, (6) HPM-SDS-PAE adhesive, (7) HPM-urea-PAE adhesive.

3.5. Viscosity Measurement

Viscosity is an important physical property related directly to the flowability of adhesives, and the optimal viscosity of wood adhesives is between 5000 and 25000 mPa·s [24]. HPM adhesive demonstrates over-penetration (Figure 6) in the application process because of its low viscosity (4378 mPa·s) [31]. Therefore, it was unable to form a denser cured adhesive layer during the hot-press cycle, thus having a poor water resistance and bonding strength [32]. The viscosity of the HPM-papain adhesive was increased by 77.8% (7784 mPa·s), and the results indicate that papain hydrolysis broke the peptide bond of HPMP molecules and degraded the molecules into small polypeptide chains [33]. After the introduction of SDS (HPM-SDS adhesive) and urea (HPM-urea adhesive), the viscosity increased by 58.5% (6940 mPa·s) and 25.8% (5509 mPa·s), which was attributed to the reduction in distance and the enhancement of intermolecular friction caused by the destruction of intermolecular hydrogen bonds [9]. Compared with the HPM adhesive, HPM-papain adhesive, HPM-SDS adhesive and HPM-Urea adhesive, the viscosity of the HPM-PAE adhesive, HPM-papain-PAE adhesive, HPM-SDS-PAE adhesive, and HPM-urea-PAE adhesive was decreased by 26.3% (3227 mPa·s), 16.7% (6484 mPa·s), 17.8% (5708 mPa·s) and 6.3% (5162 mPa·s), respectively after adding PAE. There were three reasons for this: (1) the PAE reduced the solid content of adhesives; (2) the positively charged groups of PAE form electrostatic interaction with charged protein chains, resulting in the reduction in attraction and repulsion between surrounding molecules, thus reducing the viscosity [12]; and (3) in the process of crosslinking, PAE could be embedded into molecules and act as "lubricant" [34].

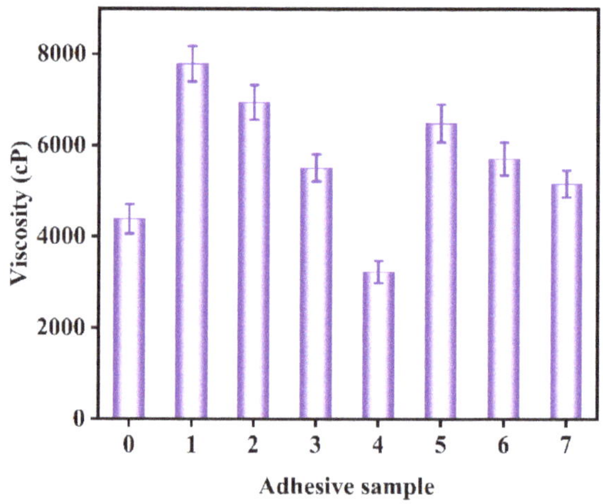

Figure 6. The viscosity of the different adhesive samples: (0) HPM adhesive, (1) HPM-papain adhesive, (2) HPM-sodium dodecyl sulfate (SDS) adhesive, (3) HPM-urea adhesive, (4) HPM-PAE adhesive, (5) HPM-papain-PAE adhesive, (6) HPM-SDS-PAE adhesive, (7) HPM-urea-PAE adhesive.

3.6. SEM Analysis

Figure 7 shows the fracture surface of the cured adhesive samples. There were a lot of holes and cracks on the surface of the HPM adhesive, which could be used as a channel for water intrusion, resulting in poor water resistance [35]. The above holes and cracks increased after the introduction of papain, indicating that the water resistance of HPMP molecules reduced, degraded by papain. This result was supported by ML and MUV. Although SDS and urea contribute to HPMP molecular unfolding, they would reduce its cohesion, so there were still holes and cracks on the surface of the HPM-SDS adhesive and HPM-urea adhesive. The massive micro-cracks appear on the HPM-PAE adhesive, indicating that the layer was brittle and could easily expose the hydrophobic pathway along the cracks [26]. This was compared with the HPM-papain-PAE adhesive, HPM-SDS-PAE adhesive, and the HPM-urea-PAE adhesive, in which the fracture surface became more compact. The results show that the denaturant treatment was helpful to further improve the crosslinking density of the network, which contributed to the water resistance and thermal behavior of the adhesive.

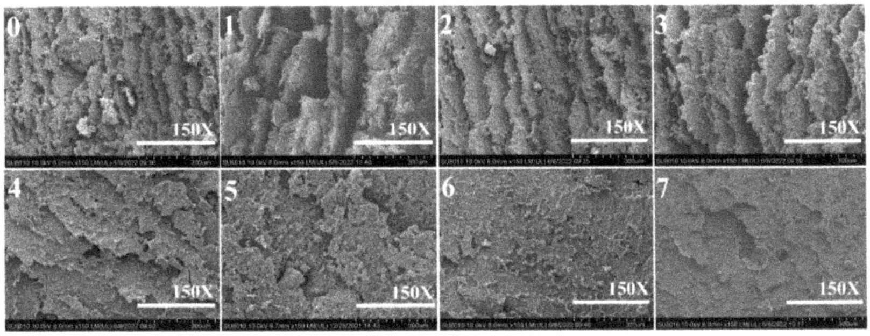

Figure 7. Fracture surface micrographs of the different cured adhesive samples: (0) HPM adhesive, (1) HPM-papain adhesive, (2) HPM-sodium dodecyl sulfate (SDS) adhesive, (3) HPM-urea adhesive, (4) HPM-PAE adhesive, (5) HPM-papain-PAE adhesive, (6) HPM-SDS-PAE adhesive, (7) HPM-urea-PAE adhesive.

4. Conclusions

In this study, (1) after papain treatment, the WS of the HPM-papain adhesive was reduced by 40.4% (0.34 MPa), the ML and MUV were increased by 34.7% (73.25%) and 38.4% (12.8%), and the viscosity increased by 141.2% (7784 mPa·s). With the addition of PAE, the WS of HPM-papain-PAE adhesive was increased 57.9% (0.9 MPa), the ML and MUV were reduced by 28.9% (38.67%) and 46.6% (4.17%), and the viscosity was decreased by 16.7% (6484 mPa·s). (2) The HPMP molecule hydrogen bond was destroyed under the SDS and urea, the water resistance was improved. Compared with pure HPM adhesive, PAE significantly improved the WS, water resistance and viscosity of the adhesive. The WS of the HPM-SDS-PAE adhesive and HPM-urea-PAE adhesive were increased by 96.4% (1.12 MPa) and 93% (1.1 MPa), the ML was reduced by 41.4% (31.88%) and 36.7% (34.44%), the MUV also showed the same trends (2.83% and 3.71%), and the viscosity was increased by 30.4% (5708 mPa·s) and 17.9% (5162 mPa·s). (3) SDS and urea could improve the water resistance. On this basis, the addition of the PAE improved the bonding strength and water resistance of the prepared plywood, which was due to the cross-linking structure formed during curing and the nail structure formed by the adhesive penetrating into the wood pores.

Author Contributions: Data curation and writing—original draft, Y.Q.; Formal analysis, Q.G.; Methodology, T.L.; Supervision and Writing—review and editing, H.L.; Supervision, Q.W. All authors have read and agreed to the published version of the manuscript.

Funding: This research was funded by the Key Research and Development Plan of Xinjiang Autonomous Region (2021B02003-3, 2021B02003-4), Subei science and technology special fund (XZ-SZ202036), Science and Technology Innovation project of Chinese Academy of Agricultural Sciences (CAAS-ASTIP-201X-IAPPST).

Institutional Review Board Statement: Not applicable.

Informed Consent Statement: Not applicable.

Data Availability Statement: Not applicable.

Conflicts of Interest: The authors declare no conflict of interest.

References

1. Gu, W.; Li, F.; Liu, X.; Gao, Q.; Gong, S.; Li, J.; Shi, S.Q. Borate chemistry inspired by cell walls converts soy protein into high-strength, antibacterial, flame-retardant adhesive. *Green Chem.* **2020**, *22*, 1319–1928. [CrossRef]
2. Sun, Z.; Sun, B.; Bai, Y.; Gao, Z. Economical improvement on the performances of a soybean flour-based adhesive for wood composites via montmorillonite hybridization. *Compos. B. Eng.* **2021**, *217*, 108920. [CrossRef]

3. He, C.; Hu, Y.; Wang, Y.; Liao, Y.; Xiong, H.; Selomulya, C.; Hu, J.; Zhao, Q. Complete waste recycling strategies for improving the accessibility of rice protein films. *Gree Chem.* **2020**, *22*, 490–503. [CrossRef]
4. Ai, X.; Feng, S.; Shui, T.; Kakkar, H.; Xu, C.C. Effects of alcell lignin methylolation and lignin adding stage on lignin-based phenolic adhesives. *Molecules* **2021**, *26*, 6762. [CrossRef] [PubMed]
5. Watcharakitti, J.; Win, E.E.; Nimnuan, J.; Smith, S.M. Modified starch-based adhesives: A review. *Polymers* **2022**, *14*, 2023. [CrossRef] [PubMed]
6. Cai, L.; Li, Y.; Lin, X.; Chen, H.; Gao, Q.; Li, J. High-performance adhesives formulated from soy protein isolate and bio-based material hybrid for plywood production. *J. Clean. Prod.* **2022**, *353*, 131587. [CrossRef]
7. Chen, Y.; Shi, A.; Wang, Q. Peanut meal as plywood adhesives: Preparation and characterization. *J. Adhes. Sci. Technol.* **2018**, *32*, 2450–2463. [CrossRef]
8. Qu, Y.; Guo, Q.; Li, T.; Zhang, S.; Wang, B.; Yue, H.; Liu, H.; Yang, J.; Wang, Q. Preparation and characterization of birch plywood prepared by hot-pressed peanut meal adhesive. *Int. J. Adhes. Adhes.* **2022**, *117*, 103186. [CrossRef]
9. Li, J.; Li, X.; Li, J.; Gao, Q. Investigating the use of peanut meal: A potential new resource for wood adhesives. *RSC Adv.* **2015**, *5*, 80136–80141. [CrossRef]
10. Chen, C.; Chen, F.; Liu, B.; Du, Y.; Liu, K. Peanut meal-based wood adhesives enhanced by urea and epichlorohydrin. *R. Soc. Open Sci.* **2019**, *6*, 191154. [CrossRef] [PubMed]
11. Bai, M.; Cao, J.; Li, J.; Li, C. Development of soybeans starch based tough, water resistant and mildew-proof adhesives through multiple cross linking cooperation strategy. *J. Clean. Prod.* **2021**, *321*, 129001. [CrossRef]
12. Liu, H.; Bean, S.; Sun, X.S. Camelina protein adhesives enhanced by polyelectrolyte interaction for plywood applications. *Ind. Crops Prod.* **2018**, *124*, 343–352. [CrossRef]
13. Li, K.; Jin, S.; Li, X.; Li, J.; Shi, S.Q.; Li, J. Bioinspired interface engineering of soybean meal-based adhesive incorporated with biomineralized cellulose nanofibrils and a functional aminoclay. *Chem. Eng. J.* **2021**, *421*, 129820. [CrossRef]
14. Zhang, Y.; Liu, Z.; Xu, Y.; Li, J.; Shi, S.Q.; Li, J.; Gao, Q. High performance and multifunctional protein-based adhesive produced via phenol-amine chemistry and mineral reinforcement strategy inspired by arthropod cuticles. *Chem. Eng. J.* **2021**, *426*, 130852. [CrossRef]
15. Wang, Z.; Zhao, S.; Kang, H.; Zhang, W.; Zhang, S.; Li, J. Mussel byssus-inspired engineering of synergistic nanointerfacial interactions as sacrificial bonds into carbon nanotube-reinforced soy protein/nanofibrillated cellulose nanocomposites: Versatile mechanical enhancement. *Appl. Surf. Sci.* **2018**, *434*, 1086–1100. [CrossRef]
16. Xu, Y.; Xu, Y.; Han, Y.; Chen, M.; Zhang, W.; Gao, Q.; Li, J. The effect of enzymolysis on performance of soy protein-based adhesive. *Molecules* **2018**, *23*, 2752. [CrossRef]
17. Wei, Y.; Yao, J.; Shao, Z.; Chen, X. Water-resistant zein-based adhesives. *ACS Sustain. Chem. Eng.* **2020**, *8*, 7668–7679. [CrossRef]
18. Wei, X.; Wang, X.; Li, Y.; Ma, Y. Properties of a new renewable sesame protein adhesive modified by urea in the absence and presence of zinc oxide. *RSC Adv.* **2017**, *7*, 46388–46394. [CrossRef]
19. *China National Standard GB/T 9846*; Plywood for General Use. Standardization Administration of the Peoples Republic of China: Beijing, China, 2015.
20. Xu, Y.; Han, Y.; Li, Y.; Li, J.; Li, J.; Gao, Q. Preparation of a strong, mildew-resistant, and flame-retardant biomimetic multifunctional soy protein adhesive via the construction of an organic-inorganic hybrid multiple-bonding structure. *Chem. Eng. J.* **2022**, *437*, 135437. [CrossRef]
21. Qu, Y.; Guo, Q.; Li, T.; Zhang, Y.; Gao, Q.; Liu, H.; Wang, Q. A novel environmentally friendly hot-pressed peanut meal protein adhesive. *J. Clean. Prod.* **2021**, *327*, 129473. [CrossRef]
22. Yue, L.; Meng, Z.; Yi, Z.; Gao, Q.; Mao, A.; Li, J. Effects of different denaturants on properties and performance of soy protein-based adhesive. *Polymers* **2019**, *11*, 1262. [CrossRef] [PubMed]
23. Zhao, S.; Wang, Z.; Pang, H.; Li, Z.; Zhang, W.; Zhang, S.; Li, J.; Li, L. Designing biomimetic microphase-separated motifs to construct mechanically robust plant protein resin with improved water-resistant performance. *Macromol. Mater. Eng.* **2020**, *305*, 1900462. [CrossRef]
24. Xu, Y.; Han, Y.; Chen, M.; Luo, J.; Shi, S.Q.; Li, J.; Gao, Q. Constructing a triple network structure to prepare strong, tough, and mildew resistant soy protein adhesive. *Compos. B. Eng.* **2021**, *211*, 108677. [CrossRef]
25. Jin, S.; Li, K.; Gao, Q.; Zhang, W.; Chen, H.; Li, J.; Shi, S.Q. Multiple crosslinking strategy to achieve high bonding strength and antibacterial properties of double-network soy adhesive. *J. Clean. Prod.* **2020**, *254*, 120143. [CrossRef]
26. Zhang, J.; Zhang, M.; Zhang, Y.; Shi, S.Q.; Gao, Q. Improving bond performance and reducing cross-linker dosage for soy flour adhesives inspired by spider silk. *ACS Sustain. Chem. Eng.* **2020**, *9*, 168–179. [CrossRef]
27. Zeng, G.; Zhou, Y.; Liang, Y.; Zhang, F.; Luo, J.; Li, J.; Fang, Z. A hair fiber inspired bio-based adhesive with high bonding strength and mildew tolerance. *Chem. Eng. J.* **2022**, *434*, 134632. [CrossRef]
28. Li, Z.; Zhao, S.; Wang, Z.; Zhang, S.; Li, J. Biomimetic water-in-oil water/pMDI emulsion as an excellent ecofriendly adhesive for bonding wood-based composites. *J. Hazard. Mater.* **2020**, *396*, 122722. [CrossRef]
29. Luo, J.; Zhou, Y.; Gao, Q.; Li, J.; Yan, N. From wastes to functions: A new soybean meal and bark-based adhesive. *ACS Sustain. Chem. Eng.* **2020**, *8*, 10767–10773. [CrossRef]
30. Chen, M.; Zhang, Y.; Li, Y.; Shi, S.Q.; Li, J.; Gao, Q.; Guo, H. Soybean meal-based wood adhesive enhanced by phenol hydroxymethylated tannin oligomer for exterior use. *Polymers* **2020**, *12*, 758. [CrossRef]

31. Luo, J.; Zhou, Y.; Zhang, Y.; Gao, Q.; Li, J. An eco-effective soybean meal-based adhesive enhanced with diglycidyl resorcinol ether. *Polymers* **2020**, *12*, 954. [CrossRef]
32. Li, J.; Luo, J.; Li, X.; Yi, Z.; Gao, Q.; Li, J. Soybean meal-based wood adhesive enhanced by ethylene glycol diglycidyl ether and diethylenetriamine. *Ind. Crops Prod.* **2015**, *74*, 613–618. [CrossRef]
33. Xu, Y.; Han, Y.; Shi, S.Q.; Gao, Q.; Li, J. Preparation of a moderate viscosity, high performance and adequately-stabilized soy protein-based adhesive via recombination of protein molecules. *J. Clean. Prod.* **2020**, *255*, 120303. [CrossRef]
34. Yuan, C.; Chen, M.; Luo, J.; Li, X.; Gao, Q.; Li, J. A novel water-based process produces eco-friendly bio-adhesive made from green cross-linked soybean soluble polysaccharide and soy protein. *Carbohydr. Polym.* **2017**, *169*, 417–425. [CrossRef] [PubMed]
35. Pang, H.; Wang, Y.; Chang, Z.; Xia, C.; Huang, Z. Soy meal adhesive with high strength and water resistance via carboxymethylated wood fiber-induced crosslinking. *Cellulose* **2021**, *28*, 3569–3584. [CrossRef]

molecules

Article

Esterified Soy Proteins with Enhanced Antibacterial Properties for the Stabilization of Nano-Emulsions under Acidic Conditions

Tingyu Wang [1,†], Kehan Yi [2,†], Yang Li [2,3], Huan Wang [3], Zhijun Fan [4], Hua Jin [1,*] and Jing Xu [1,*]

1 College of Arts and Sciences, Northeast Agricultural University, Harbin 150030, China; 15245972972@163.com
2 National Research Center of Soybean Engineering and Technology, Harbin 150028, China; yikehan@neau.edu.cn (K.Y.)
3 College of Food Science, Northeast Agricultural University, Harbin 150030, China; whname@neau.edu.cn
4 Heilongjiang Beidahuang Green and Healthy Food Co., Ltd., Jiamusi 154007, China; 15845177666@139.com
* Correspondence: jinhua@neau.edu.cn (H.J.); xujing@neau.edu.cn (J.X.)
† These authors contributed equally to this work and should be considered co-first authors.

Abstract: Soy protein isolate (SPI), including β-conglycinin (7S) and glycinin (11S), generally have low solubility under weakly acidic conditions due to the pH closed to their isoelectric points (pIs), which has limited their application in acidic emulsions. Changing protein pI through modification by esterification could be a feasible way to solve this problem. This study aimed to obtain stable nano-emulsion with antibacterial properties under weakly acidic conditions by changing the pI of soy protein emulsifiers. Herein, the esterified soy protein isolate (MSPI), esterified β-conglycinin (M7S), and esterified glycinin (M11S) proteins were prepared. Then, pI, turbidimetric titration, Fourier transform infrared (FTIR) spectra, intrinsic fluorescence spectra, and emulsifying capacity of esterified protein were discussed. The droplet size, the ζ-potential, the stability, and the antibacterial properties of the esterified protein nano-emulsion were analyzed. The results revealed that the esterified proteins MSPI, M7S, and M11S had pIs, which were measured by ζ-potentials, as pH 10.4, 10.3, and 9.0, respectively, as compared to native proteins. All esterified-protein nano-emulsion samples showed a small mean particle size and good stability under weakly acidic conditions (pH 5.0), which was near the original pI of the soy protein. Moreover, the antibacterial experiments showed that the esterified protein-based nano-emulsion had an inhibitory effect on bacteria at pH 5.0.

Keywords: esterification; soy protein; nano-emulsion; acidic; antibacterial

check for updates

Citation: Wang, T.; Yi, K.; Li, Y.; Wang, H.; Fan, Z.; Jin, H.; Xu, J. Esterified Soy Proteins with Enhanced Antibacterial Properties for the Stabilization of Nano-Emulsions under Acidic Conditions. *Molecules* **2023**, *28*, 3078. https://doi.org/10.3390/molecules28073078

Academic Editors: Shudong He, Wei Xu and Muhammad H. Aludatt

Received: 13 March 2023
Revised: 26 March 2023
Accepted: 27 March 2023
Published: 30 March 2023

1. Introduction

Generally, nano-emulsions are defined as colloidal systems with a particle size in the range of 50–500 nm that exhibit greater resistance to creaming, sedimentation, coalescence, and flocculation than conventional emulsions, thus exhibiting higher stability [1]. When emulsions are used, they often involve acidic environments (pH 4.0–7.0) [2]. Therefore, whether the emulsion is stable at an acidic pH is a matter of concern. In particular, the improvement of nano-emulsion applications is closely related to the utilization of emulsifiers. As a kind of natural emulsifier with a wide range of sources and select functional properties, soybean protein has been widely used in the preparation of emulsion—based systems [3,4]. However, the solubility and emulsifying properties of soy protein in weakly acidic environments has typically been low, because when the pH value is close to the isoelectric point (pI) of the protein (pI $\approx$ 4.5), the charge of the protein molecule is almost zero, and the intermolecular repulsion is weak [5], which is not conducive to the application of nano-emulsions made of soybean protein.

Until recently, changing the protein pI by chemical modification has been considered an effective method to improve the solubility of protein near the original pI [6–8]. The esterification reaction has blocked the carboxyl groups of the protein, so that the net positive charge of the protein increases, along with the pI, thereby causing a higher solubility of the protein in a low-pH range. Sitohy et al. [9] reported that the pI of esterified soy protein could be increased from 4.5 to 8.0. This was attributed to the formation of esters between the alcohol and carboxyl groups, reducing the content of negatively charged groups on the protein surface.

In addition, esterification is also a feasible way to ensure that a protein achieves antibacterial activity [9,10]. Generally, a protein is positively charged when pH < pI [11]. Blocking the free carboxyl groups (mainly derived from aspartic acid and glutamic acid) of the protein by alcohol ensures that the protein carries positive charges in a wider pH range. The positive charge surface of a protein was then able to act on the negatively charged cell membrane in order to play a role in inhibiting the growth of bacteria [9,12]. Wang et al. [13] found that lactoferrin, after esterification, carried positive charges and had a direct fungitoxic property against the pathogens of blue mold in apple fruit. Sitohy et al. [9] found that esterified legume proteins had a good inhibitory effect on Gram-positive and Gram-negative bacteria due to the presence of static electricity. In conclusion, the esterified proteins exhibited excellent solubility, emulsification, and antibacterial properties under weakly acidic conditions. An emulsion prepared by esterified proteins as an emulsifier had smaller particle sizes, more uniform internal structures, and stronger stability. Therefore, it should be valuable to apply esterification as a modification to improve not only the functional properties of the protein or the preparation of stable emulsions at low pH values but also the antibacterial properties of proteins. However, little is known about the preparation of esterified protein nano-emulsions and their antibacterial properties, so more research is required to explore this process.

In this study, oil-in-water (O/W)-type nano-emulsions (MSPI-NE, M11S-NE, and M7S-NE) were prepared by ethanol-esterified soybean proteins (MSPI, M11S, and M7S). The structures and properties of the esterified proteins were discussed, and the stability and the antibacterial properties of emulsion were also analyzed under weakly acidic conditions. This work provided theoretical proof that soybean proteins could prepare nano-emulsions under weakly acidic conditions and had the dual functions of emulsifier and antiseptic, broadening the prospects for its application in the food-processing, pharmaceutical, and cosmetic industries.

2. Results and Discussion

2.1. Protein Isoelectric Point (pI)

The pI value of protein is often affected by chemical side-chain modifications, the amino-acid composition, and the molecular conformation [14]. The surface charge of the proteins was measured to reveal the pI of the esterified proteins. As shown in Figure 1b, the pI of M11S occurred at a pH of 9.0, according to the zero ζ-potential of M11S at this pH. For the same reason, the pIs of MSPI and M7S occurred at a pH of 10.3 and 10.4, respectively. Therefore, the pI orders of MSPI, M11S, and M7S were as follows: M7S > MSPI > M11S, and the increases were consistent with the orders of the esterification rates. After the reaction between the protein carboxyl groups and ethanol, significantly, the pI of the ethanol-esterified protein had increased due to the negatively charged carboxyl group of the protein being blocked [9]. As shown in Figure 1a, similar results were observed in the esterified soybean, broad-bean, and chickpea protein isolates (pI $\approx$ 8.0) [9]. Moreover, the lowest pI of M11S could have been related to the compact conformation of 11S linked by an acid such as A polypeptide (37–42 kDa) and a basic B polypeptide (17–20 kDa) via the disulfide bonds [15]. Therefore, it could have had a large number of carboxyl groups embedded in the protein molecule, reducing the number of the carboxyl groups that could react with the ethanol. However, esterification was still an effective method for increasing the positive charges carried by proteins in a wider pH range, especially for SPI and 7S.

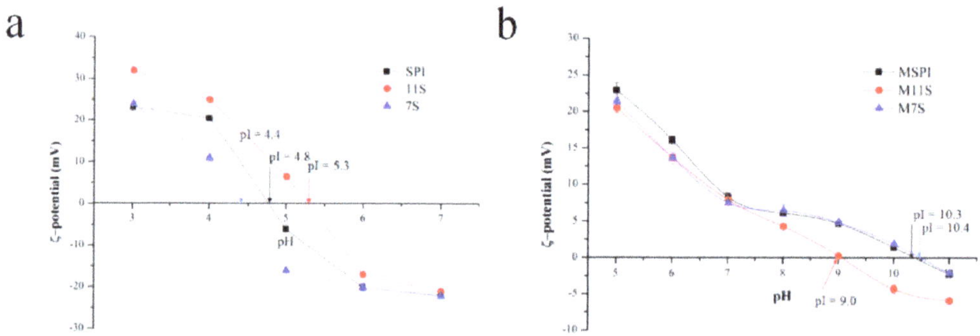

Figure 1. ζ-potetial (mV) of native protein (**a**) and esterified protein (**b**) solution at different pH.

2.2. *Turbidity*

The turbidity analysis is an intuitive technique to characterize the number and the size of the complex aggregates of proteins [16]. The magnitude of the increase in turbidity depends on the number, the size, and the refractive index of the contrast of the particles [17]. Generally, larger sizes of protein particles dispersed in a solution lead to higher absorbance measurements, which indicate a higher turbidity value. As shown in Figure 2, the turbidity measurements of the native SPI, 11S, and 7S under a range of 4.0–5.5 pH were significantly greater than at other pH values, and this phenomenon revealed the formation of the larger protein particles via pH-induced self-assembly due to their low ζ-potentials and weak electrostatic repulsion [18]. In the range of 3.0–7.0 pH, the turbidity values of the esterified protein solutions were very low, indicating that these were stable solutions. As the pH continued to increase, the turbidity values of all the esterified protein solutions increased significantly, indicating that the esterified proteins did not aggregate at acidic pH values but at alkaline pH values. This verified the results of the pI: The modification by esterification had increased the positive charge of the protein, and thus, the pI had increased.

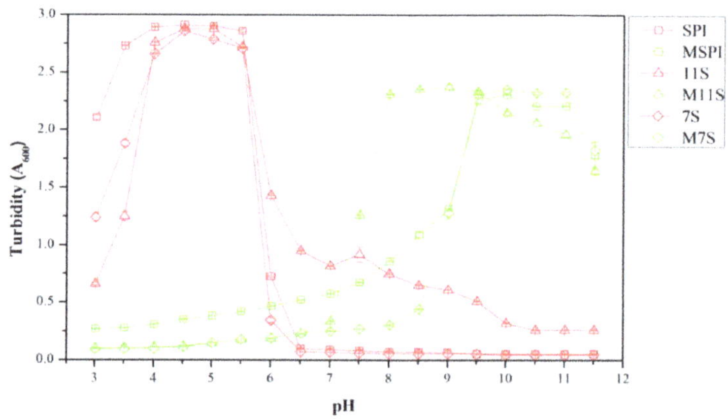

Figure 2. Turbidity curve of native/esterified protein solutions from 3.0 to 11.5 pH.

2.3. *FTIR Spectra*

To determine the formation of the ester groups in the proteins, the infrared spectrum of the esterified proteins, and the control were measured. The FTIR results of the proteins are shown in Figure 3. A typical broad peak around 3200–3400 cm^{-1} was shown in all the samples due to the intermolecular H–bond and the stretching vibration of O–H and N–H [19]. The absorption peak around 2960 cm^{-1} was attributed to the C–H stretching vibration of the CH_3 and CH_2 groups in the protein. The peaks around 1650 and 1540 cm^{-1} in the FTIR spectra of the native proteins were assigned to the C=O stretching vibration and C–N stretching vibration of the amide-I band and the N–H deformation vibration and the stretching vibration of C–N in the amide-II band. At the same time, the esters had 2 characteristically strong absorption bands as a result of the C=O (1750–1735 cm^{-1}) and C–O (1300–1000 cm^{-1}) stretching [20]. All the esterified proteins (MSPI, M11S, and M7S) showed a small peak at 1733 cm^{-1} due to the C=O absorption band, indicating the formation of the ester groups in the proteins (Figure 3a–c). The results indicated that the esterified proteins were successfully prepared, as these were similar to the results obtained by Wang et al. [20].

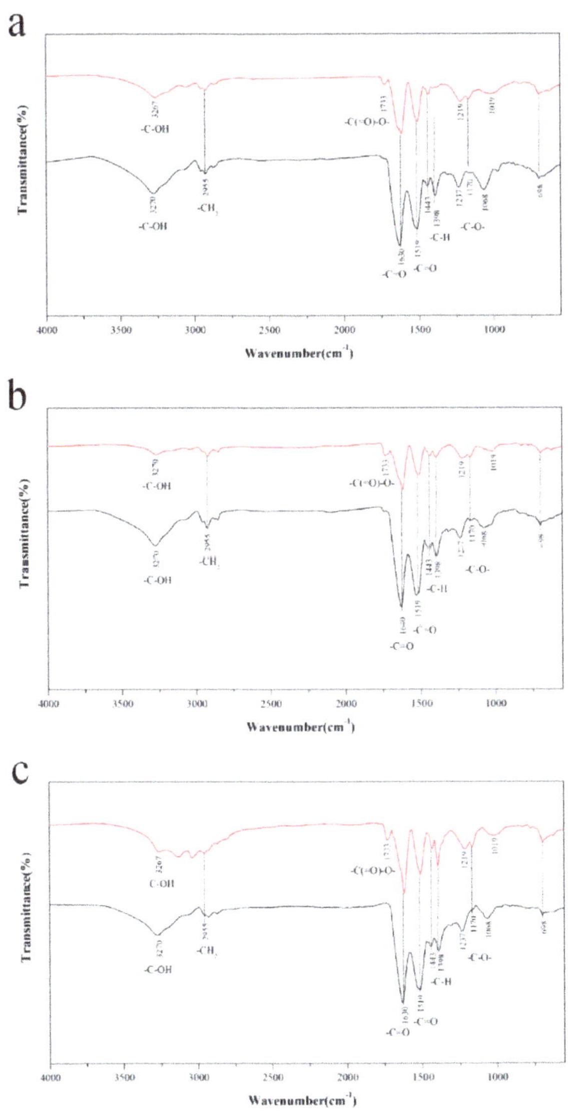

Figure 3. (**a**) FTIR spectra of MSPI (red line) and SPI (black line); (**b**) FTIR spectra of M11S (red line) and 11S (black line); (**c**) FTIR spectra of M7S (red line) and 7S (black line).

2.4. Fluorescence Spectra

The structural and polarity information of the proteins was investigated by the intrinsic fluorescence spectra [21]. Generally, the change in the protein conformation in the surrounding environment of the tryptophan residues would be reflected by the changes in the tryptophan fluorescence intensity and the shift trends, depending on whether the tryptophan residues were exposed or buried [22]. As shown in Figure 4a–c, the maximum emission intensity of the protein fluorescence spectrum was greatly affected by the pH. The fluorescence intensity of the native proteins gradually decreased as the pH decreased from 9.0 to 5.0. This phenomenon was attributed to the tryptophan being masked when the native soy proteins aggregated at pH 5.0, as it was then closed to the pI [23]. Similarly, due to the higher pIs of the esterified proteins, MSPI, M11S, and M7S had the lowest fluorescence intensity at pH 9.0. As compared to the native proteins, the higher fluorescence intensity values for the esterified proteins at all the pH values suggested that the tryptophan residues were in the slightly unfolded protein structures of MSPI, M11S, and M7S [23].

2.5. Emulsifying Activity Index (EAI) and Emulsion Stability Index (ESI)

Native soy protein is a multifunctional protein, which has greatly restricted its application in weakly acidic fields due to its poor water solubility near its pI. In this study, the emulsifying abilities of the esterified proteins were evaluated to analyze the probability of the fabrication of an esterified protein emulsion under weakly acidic conditions. As shown in Figure 5, the emulsifying properties of an esterified protein at 5.0 pH was satisfactory after esterification, which had changed the pI and the solubility, which was similar to the results of Sitohy et al. [10] on esterified β-lactoglobulin. As the pH increased, the EAI values of the esterified soybean protein decreased. This result was attributed to the fewer charges and the worsened solubility of the esterified protein emulsifiers near pI, as the esterified proteins could not fully be absorbed on the surface of the droplets or supply enough electrostatic repulsion between the oil droplets. Until they had reached a pH of 9.0, the emulsifying abilities of the esterified proteins were very low, and the EAI values of MSPI, M11S, and M7S were 3.6, 1.4, and 4.9 m^2/g, respectively. Moreover, from Figure 5b, we observed that the ESI values of the esterified proteins showed the same trend as EAI. Therefore, the esterification had enhanced the emulsification of soy protein at 5.0 pH, laying a theoretical foundation for the preparation of a stable and uniform emulsion system under weakly acidic conditions. Then, the pHs of 5.0 and 7.0 were selected for subsequent research to further analyze the application potential of the esterified proteins for emulsion preparation.

2.6. Physical Properties of Emulsions

2.6.1. Particle Size and ζ-Potential of Emulsions

The nano-emulsions were prepared using esterified proteins at pHs of 5.0 and 7.0, and the controls were the emulsions prepared with the corresponding natural proteins. The average particle size and ζ-potential of the different emulsion samples are shown in Table 1. The emulsions prepared from the native proteins were unstable at a pH of 5.0 and separated quickly (Figure 6), with average particle sizes in the micrometer range. However, the emulsions prepared by the esterified proteins were stable and uniformly dispersed at a pH of 5.0 with a small particle size. The average particle diameters of MSPI-NE, M11S-NE, and M7S-NE were 256.9 nm, 438.7 nm, and 257.0 nm, respectively. This difference was consistent with the result of the EAI and ESI. At a pH of 5.0, the ζ-potentials of the emulsions stabilized by MSPI, M11S, and M7S were +26.7, +22.7, and +27.1 mV (Table 1), respectively. They were much higher than the ζ-potentials of +11.2, +6.3, and +11.9 mV, respectively, at a pH of 7.0.The ζ-potentials of the emulsions stabilized by MSPI, M11S, and M7S were +11.2, +6.3, +11.9 mV, respectively, at a pH of 7.0. Therefore, at a pH of 7.0, the mean particle sizes of all the esterified protein nano-emulsions were larger than those at a pH of 5.0. This was attributed by the weakening of the electrostatic repulsion on the surface of the oil droplets, and the oil droplets tended to form larger droplets. There were no significant differences

($p > 0.05$) between the particle sizes of MSPI-NE, as compared to SPI-NE, M7S-NE, and 7S-NE, at a pH of 7.0. This illustrated that the esterified proteins could still stabilize emulsions under neutral conditions because the ζ-potentials of MSPI-NE and M7S-NE were 11.2 and 11.9 mV, respectively, which could still provide a certain amount of electrostatic repulsion. In addition, the surface hydrophobicity of the soy proteins had been reported to increase after modification by esterification [24,25]. As a result, the modification by esterification enabled the soy proteins to be more easily adsorbed at the oil-water interface and form a sturdy interfacial layer, thereby stabilizing the emulsion with a smaller droplet size via strong steric resistance [26].

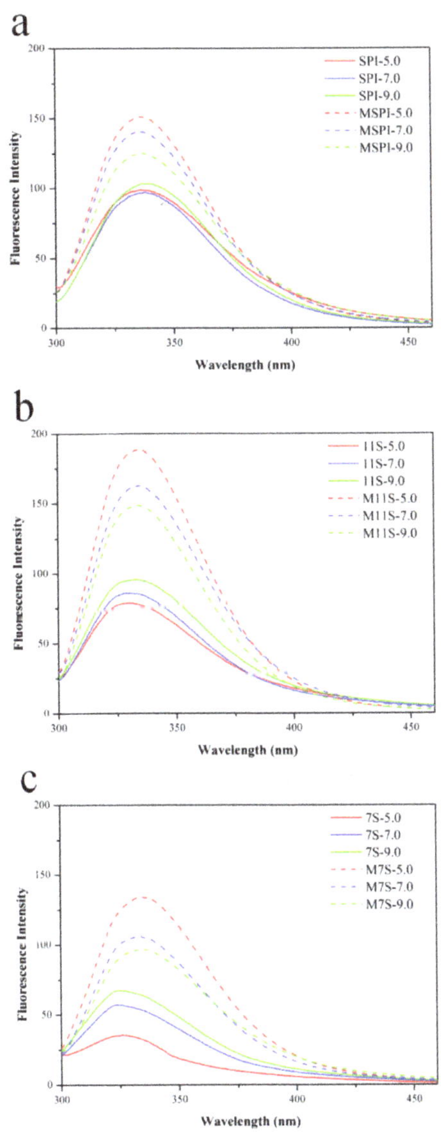

Figure 4. Intrinsic fluorescence at different pH values of MSPI and SPI (**a**); M11S and 11S (**b**); M7S and 7S (**c**).

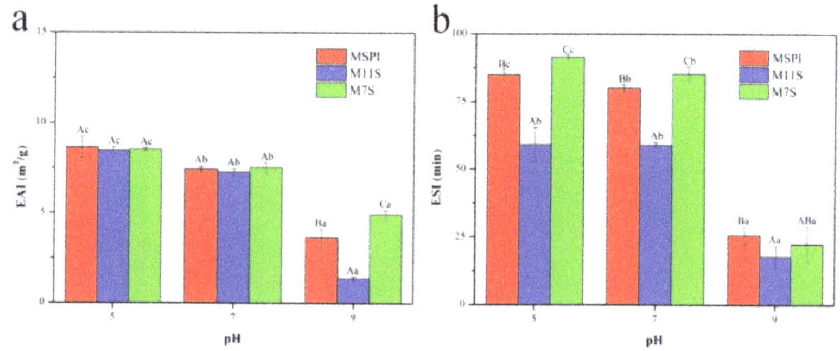

Figure 5. Emulsifying activity index (EAI) (**a**) and emulsion stability index (ESI) (**b**) for esterified soy proteins under different pH conditions. Different capital letters (A–C) indicate that under the same pH conditions, there were significant differences in EAI/ESI results among the different proteins ($p < 0.05$), and different lowercase letters (a–c) indicate that the same protein sample had significant differences in their EAI/ESI values at different pHs ($p < 0.05$).

Table 1. Mean particle sizes (nm) and ζ-potentials (mV) of oil-in-water nano-emulsions stabilized by various proteins at different pH values.

Sample	Mean Particle Size (nm)		ζ−Potential (mV)	
	pH 5.0	pH 7.0	pH 5.0	pH 7.0
SPI-NE	3500 ± 100^{c}	278.2 ± 5.5^{a}	-1.7 ± 0.1^{b}	-33.2 ± 1.8^{a}
MSPI-NE	256.9 ± 10.1^{a}	278.4 ± 3.0^{a}	26.7 ± 0.5^{e}	11.2 ± 0.3^{d}
11S-NE	5130 ± 170^{e}	357.8 ± 8.2^{b}	1.2 ± 0.1^{c}	-30.6 ± 0.5^{b}
M11S-NE	438.7 ± 17.2^{b}	479.0 ± 9.7^{c}	22.7 ± 0.2^{d}	6.3 ± 0.1^{c}
7S-NE	3840 ± 140^{d}	273.6 ± 8.4^{a}	-1.4 ± 0.1^{a}	-32.6 ± 0.7^{a}
M7S-NE	257.0 ± 8.1^{a}	270.8 ± 7.5^{a}	27.1 ± 0.1^{e}	11.9 ± 0.6^{d}

Data not sharing the same superscript letter in the same column are significantly different ($p < 0.05$).

Figure 6. Appearance of oil-in-water emulsions prepared at different pH for native and esterified proteins.

2.6.2. Emulsion Morphology

The microstructures of the esterified protein nano-emulsions were further observed by TEM. First, the images of the microstructures of the native protein emulsions were acquired at a pH of 5.0 (Figure 7a–c). The TEM images showed the flocculation of the oil droplets in the fresh emulsions were stabilized by native proteins, as proteins typically precipitated near the pI [27]. However, the oil droplets of the esterified protein nano-emulsions maintained better spherical microstructures, which was a result of the relatively high electrostatic repulsion and the space resistance between the droplets (Figure 7d–f). At a pH of 7.0, the native protein-based nano-emulsions (Figure 7g–i) and the esterified protein-based nano-emulsions (Figure 7j–l) all displayed regular spherical particles. This suggested that both the esterified and native proteins could be adsorbed on the surface of the oil droplets sufficiently to maintain uniform and stable emulsions. The TEM image results were consistent with the particle-size and ζ-potential results of the emulsions.

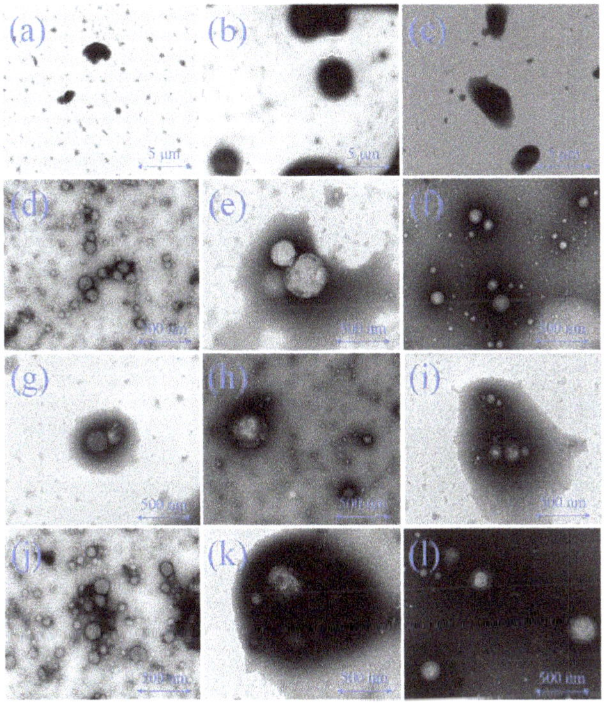

Figure 7. Transmission electron microscopy (TEM) analysis of SPI-NE (pH 5.0, (**a**)), 11S-NE (pH 5.0, (**b**)), 7S-NE (pH 5.0, (**c**)), MSPI-NE (pH 5.0, (**d**)), M11S-NE (pH 5.0, (**e**)), M7S-NE (pH 5.0, (**f**)), SPI-NE (pH 7.0, (**g**)), 11S-NE (pH 7.0, (**h**)), 7S-NE (pH 7.0, (**i**)), MSPI-NE (pH 7.0, (**j**)), M11S-NE (pH 7.0, (**k**)), M7S-NE (pH 7.0, (**l**)).

2.7. Emulsion Stability Analysis

It was important to examine the influence of the storage time and the temperature changes on the stability of the oil-in-water nano-emulsions that had been stabilized by the esterified proteins. This information would be essential for determining the applicable fields for the esterified protein-stabilized nano-emulsions so they could be successfully employed in the food industry.

2.7.1. Storage Stability

It is well known that long-term storage can lead to the destabilization of emulsions by flocculation, coalescence, and Ostwald ripening [28], and droplet size was a parameter that assisted in determining the kinetic stability of the emulsions during storage. As shown in Figure 8, at a pH of 5.0, the mean particle sizes had remained relatively small (<500 nm) during the 28-day storage period, although there had been an appreciable increase on the 21st day in the esterified-protein nano-emulsions (Figure 8a). The particle size results indicated that the esterified protein had the ability to maintain stable nano-emulsions under weakly acidic conditions. In addition, considering the highest esterification rates and ζ-potentials, the modification by esterification could have had the strongest effect on the surface charge and the hydrophobicity of 7S. Therefore, M7S showed the strongest ability to generate electrostatic repulsion and supply space resistance; thus, the emulsion prepared by M7S showed the best stability at the end of storage.

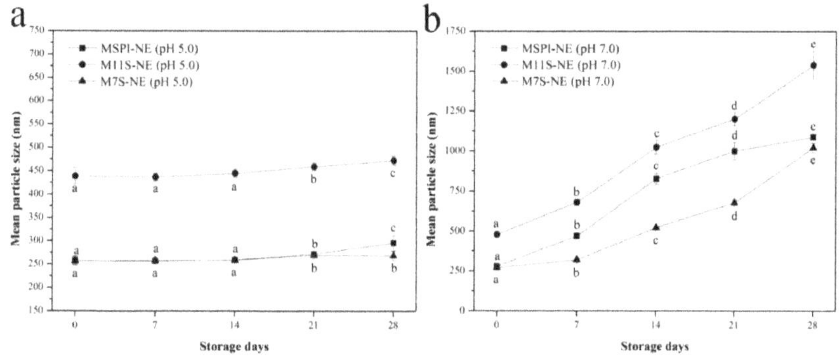

Figure 8. Mean particle sizes (nm) of nano-emulsions during the 28-day storage at 4 °C: pH 5.0 (**a**); pH 7.0 (**b**). Different lowercase letters (a–e) indicate that there are significant differences in the same emulsion after storage ($p < 0.05$).

Though the esterified proteins had prepared nano-emulsions with small particle sizes, the particle size of the emulsion increased significantly after 7 days at a pH of 7.0 (Figure 8b). There was appreciable droplet aggregation in the emulsions because of the poor electrostatic repulsion induced by the low net-charge of the droplets at a pH of 7.0 (Table 1), which could not, therefore, stabilize the emulsion in long-term storage. The relationship between the emulsion stability and the electrostatic repulsion had been confirmed in many studies. For example, Wang et al. [29] found a similar result: Emulsions stabilized by soy protein isolates complexes showed weak electrostatic repulsions, so the particle size of the emulsion had changed significantly after storage.

2.7.2. Temperature Treatment

Emulsion products may be subjected to various effects of temperature fluctuations during industrial production, such as high-temperature sterilization or low-temperature storage to extend their shelf lives. Therefore, it was important to determine the influence of temperature on the particle size of the esterified protein nano-emulsions. For the soy proteins, the temperature conditions could lead to conformational changes, exposing the interior hydrophobic patches and decreasing solubility [30,31]. The particle sizes of the emulsions after high-temperature heating and freeze-thaw treatments were measured, as shown in Figure 9. After freeze–thaw treatment, all nano-emulsions produced oil droplets with larger particle sizes by coalescence and flocculation, though to a different extent (Figure 9a,b). At a pH of 5.0, the mean particle sizes of MSPI-NE and M7S-NE increased from 256.9 nm to 293.0 nm and from 257.0 nm to 275.8 nm, respectively. Accordingly, the small increments in mean particle sizes suggested that MSPI-NE and M7S-NE had acceptable freeze-thaw stability. The phenomenon of droplet aggregation during freezing and thawing was mainly related to the destruction of the emulsifier layer by the ice crystals. The modification by esterification improved the hydrophobicity of the proteins [21], so the adsorption of the proteins at the oil-water interface became strong, contributing to the stability of the nano-emulsions. However, the droplets of M11S-NE gathered significantly (>2 μm), which was attributed to the properties of 11S that precipitates after the freezing treatment [16,32]. The 11S-protein molecules tended to aggregate at low temperatures, which caused the emulsifier layer at the oil-water interface to be insufficient for the stabilization of the nano-emulsions. In addition, during freezing, the concentration of the ions in the unfrozen water phase could increase, resulting in electrostatic shielding that promoted a decrease in the electrostatic repulsion between the droplets [33]. This could explain why the particle sizes of the nano-emulsions with a pH of 7.0 increased more significantly. Because of the relatively low net-charge of the droplets at a pH of 7.0, they were more affected by the electrostatic shielding.

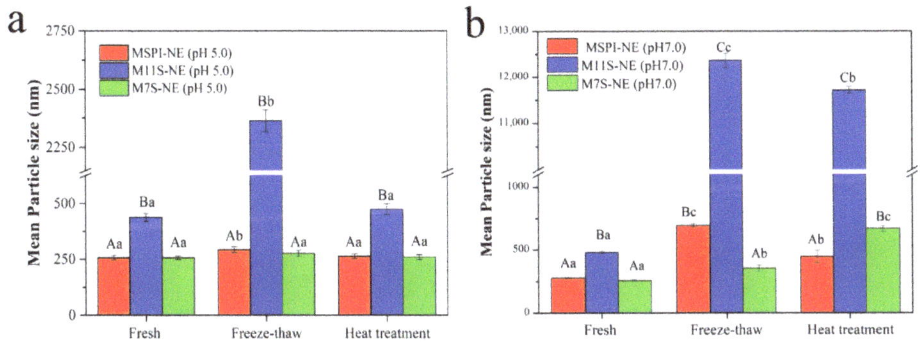

Figure 9. Mean particle size of fresh, freeze-thawed, or heat-treated nano-emulsions at a pH of 5.0 (**a**) and a pH of 7.0 (**b**); Different capital letters (A–C) indicate significant differences in mean particle size between different nano-emulsions under the same treatment ($p < 0.05$), and different lowercase letters (a–c) indicate significant differences ($p < 0.05$) in the mean particle size of the same emulsion sample under different treatments.

After being treated at 95 °C for 30 min, the mean particle sizes of MSPI-NE, M11S-NE, and M7S-NE changed from 256.9 nm, 438.7 nm, and 257.0 nm, respectively, to 263.6 nm, 475.2 nm, and 260.1 nm (Figure 9a), respectively, indicating that the effect of the heat treatment had been limited at a pH of 5.0. On the one hand, the modification by esterification improved the emulsification of the native proteins under acidic conditions, the proteins could more easily be adsorbed on the surface of the oil droplets to form compact interface layers in order to stabilize the emulsion through steric hindrance [10]. On the other hand, the charge densities of the esterified proteins were higher at a pH of 5.0, which provided sufficient electrostatic repulsion to inhibit the aggregation of the emulsion droplets. In addition, the tertiary structures of the proteins unfolded after proper heat treatments, and then the flexibility and the emulsification of the protein molecules increased, which was also beneficial for the emulsion stability [34]. Regarding the thermal stability, M11S-NE still showed the worst results among all the esterified protein nano-emulsions. According to the above experimental results, the lower positive charges on the droplet surfaces of M11S-NE (Table 1) may not have been enough to inhibit the aggregation of the droplets after the heat treatment. In summary, the three esterified proteins were effective in stable emulsions, but the modification of the densely structured 11S showed a slightly lower nano-emulsion stability than SPI and 7S.

2.8. Bacteriostatic Analysis of Nano-Emulsion

In recent years, many studies have confirmed that esterified proteins have the functions of inhibiting and killing bacteria and microorganisms due to their positive charges, which act on the negatively charged cell membrane to restrain the growth of bacteria [9,13,35]. In this experiment, the esterified soy proteins appeared to have dual functions as emulsifiers and bacteriostatic agents.

It was found that the esterified protein-based nano-emulsions had an inhibitory effect on the 3 kinds of bacteria, as shown in Figure 10, but no antibacterial activity was found in any of the control samples. This result was attributed to the fact that the esterified protein carried a positively charged amino group and the bacterial surface potential was usually negative [9]. The positively charged droplets could interact with the cell membrane electrostatically, and then, the life activities of the bacteria were disturbed. As compared to the microbial inhibition zone (Figure 10), we found that M7S-NE and MSPI-NE had a good inhibitory effect on *S. aureus* and *S. enteritidis*. The diameters of the inhibitory zone of MSPI-NE were 10.74 mm and 10.66 mm, and the diameters of the inhibitory zone of M7S-NE were 10.55 mm and 11.32 mm. The inhibitory effect of the acid emulsion on *E. coli* was lower, and the inhibitory zone of MSPI-NE, M11S-NE, and M7S-NE were 10.33 mm, 9.29 mm, and 10.38 mm, respectively. This result may have been related to the good growth trend of *E. coli*. M11S-NE had the lowest amount of positive charge, and it also had a degree of antibacterial activity against *E. coli*. Generally, the more positive the charge carried by the bacteriostatic agent, the stronger the electrostatic interaction with the cell membrane, and the easier it was to adsorb around the bacteria and destroy the morphology of the cell membrane [36,37]. Sitohy et al. [9] found that an esterified legume-protein solution had strong antibacterial properties due to the presence of a large amount of positive charges that could act on the cell membrane. However, for proteins with low rates of esterification, such as an esterified broad-bean protein (MBPI), the small electrostatic force between the protein and the cell membrane led to the weaker antibacterial activity of MBPI.

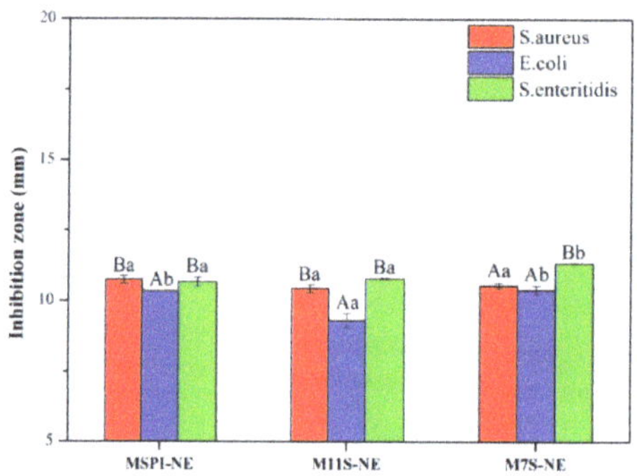

Figure 10. Diameters (mm) of the antibacterial zone formed by nano-emulsions at a pH of 5.0. For the same bacteria, values of different nano-emulsions followed by different lowercase letters mean significant differences ($p < 0.05$); for the same nano-emulsion sample, values of different bacteria followed by different capital letters mean significant differences ($p < 0.05$).

In addition, the antibacterial properties of MSPI-NE, M11S-NE, and M7S-NE were undiscovered at a pH of 7.0. As shown in Table 1, the positive charges of the nano-emulsions were high (MSPI-NE: 26.7 mV; M11S-NE: 22.7 mV; M7S-NE: 27.1 mV) at a pH of 5.0, but they were significantly decreased at a pH of 7.0 (MSPI-NE: 11.2 mV; M11S-NE: 6.3 mV; M7S-NE: 11.9 mV), which could have reduced the electrostatic interaction at a pH of 7.0 between the esterified proteins and the bacteria [38]. In summary, the esterified proteins had higher positive charged under acidic conditions, and therefore, they could inhibit the growth of bacteria, but the bacteriostasis of the proteins was weakened or even lost as the pH increased.

In the future, we will study the delivery characteristics of the esterified protein nano-emulsions further. The esterified protein nano-emulsions could be embedded with bioactive substances. And established an in vitro digestion model to investigate the digestion characteristics of the esterified protein nano-emulsions, which will be investigated at three digestion stages, including at the mouth, in the stomach, and in the small intestine.

3. Materials and Method

3.1. Materials

Low-temperature defatted soybean meal was bought from the Zhaoyuan Food Factory, Shandong, China. *Staphyloccocus aureus* (*S. aureus*), *Escherichia coli* (*E. coli*), and *Salmonella enteritidis* (*S. enteritidis*) were obtained from the China General Microbiological Culture Collection Center. Corn oil was purchased from the Yihai Kerry foodstuffs marketing company (Harbin, China). All chemicals used in this study were of analytical grade.

These were all the samples used in this study: soybean protein isolate (SPI), β-conglycinin (7S), glycinin (11S), esterified soybean protein isolate (MSPI), esterified β-conglycinin (M7S), esterified glycinin (M11S), nano-emulsion prepared by esterified soy protein isolate (MSPI-NE), nano-emulsion prepared by esterified β-conglycinin (M7S-NE), and nano-emulsion prepared by esterified glycinin (M11S-NE).

3.2. Protein Sample Preparation

The soy protein isolate (SPI), β-conglycinin (7S), and glycinin (11S) proteins were extracted according to previous research methods [39,40]. The precipitate was washed 3 times, dissolved with ultrapure water to pH 7.0, and then dialyzed for 3 days and lyophilized. The protein contents of SPI, 7S, and 11S were determined by the Kjeldahl method [12] (N × 6.25) to be 91.53 ± 1.27%, 91.74 ± 1.58%, and 93.62 ± 1.35% *w*/*w*, respectively.

3.3. Protein Esterification

Samples were prepared following the procedure of Sitohy et al. [9], with some modifications. Esterified proteins preparation is shown in Figure 11. All proteins (SPI, 7S, or 11S) were esterified at 4 °C by dispersing 5% *w*/*v* proteins in ethanol (>99.5%). The amount of aspartic acid (Asp) and glutamic acid (Glu) contents in SPI, 11S, and 7S samples were determined by amino acid analysis to calculate the amount of free carboxyl [41]. Amounts of hydrochloric acid equivalent to 50 molar ratio (acid H^+/COO^-) were added drop-wise at the beginning of the reaction to induce the protonation of the carboxylate on protein. After continuous stirring for 10 h, the samples were centrifuged at 10,000 r/min for 10 min. The esterified protein precipitate was dispersed in cold distilled water and rinsed repeatedly three times to remove residual ethanol and hydrochloric acid. Vacuum filtration was used instead of centrifugation to obtain precipitate, and then the precipitate was dissolved in distilled water at pH 7.0. Finally, the samples were dialyzed against distilled water at 4 °C for 24 h and then lyophilized. The amount of carboxyl groups that had been esterified by ethanol were determined based on the formation of colored hydroxymate–ferric ion chelate [2]. Modification was quantified based on the standard curve for the molar adsorption of iron chelate (R^2 = 0.99) that was prepared using ethyl thioacetate as a standard. In this study, the esterification rate of esterified proteins MSPI, M11S, M7S was 75.49 ± 0.36%, 69.64 ± 0.43%, and 78.16 ± 0.25%, respectively.

3.4. Isoelectric Point (PI) Determination of Esterified Proteins

The method of Wang et al. [7] was referenced and slightly modified. The native and esterified protein samples were dissolved in deionized water, and the ζ-potentials of different protein dispersions (1 mg/mL) at pH 3–11 were determined using a Zetasizer Nano ZS90 instrument (Malvern Instruments Ltd., Worcestershire, UK). A total of 0.1, 0.5, and 1 M hydrochloric acid (HCl) solutions or sodium hydroxide (NaOH) solutions were used for pH adjustment. The pI was considered to be the pH corresponding to a zero ζ-potential in the ζ-potential-versus-pH curve.

3.5. Turbidimetric Titration

The method of Dong et al. [42] was referenced and slightly modified. The pH values of native and esterified proteins (1 mg/mL) were adjusted from 3.0 to 11.5, varying by 0.5 units, and the turbidity was measured at each pH to form a turbidity titration curve.

Turbidity was expressed as the absorbances of the samples at 600 nm, which were measured using a UV-2450 spectrophotometer (Shimadzu, Kyoto, Japan).

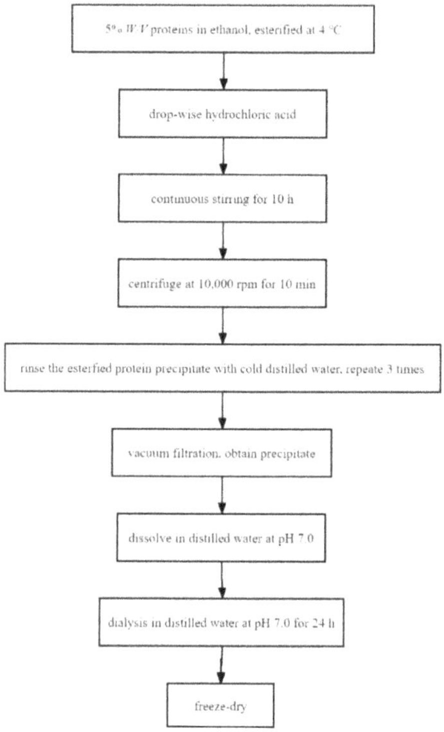

Figure 11. Flowchart of preparation of esterified protein.

3.6. Structural Analysis of Proteins

3.6.1. Fourier Transform Infrared (FTIR) Analysis

The samples were analyzed by FTIR spectrometer (IRTracer-100, Shimadzu, Kyoto, Japan), operating at 0.09 cm^{-1} resolution with 64 scans per test. Native and esterified soy protein samples (2 mg) were mixed with kalium bromatum (KBr) (200 mg) and placed under the probe, and spectroscopic data were collected from 4000 to 400 cm^{-1} in the absorbance mode [43].

3.6.2. Intrinsic Fluorescence Spectra

The intrinsic fluorescence spectra of native and esterified protein solutions (0.5 mg/mL) at pHs of 5.0, 7.0, and 9.0 were acquired using an F-4700 fluorescence spectrophotometer (Hitachi, Tokyo, Japan). The protein was excited at 290 nm, and the fluorescence intensity in the wavelength range of 300–500 nm was recorded as the emission spectrum. Both the excitation slit and the emission slit were 5 nm [44].

3.7. Emulsifying Capacity of Proteins

Emulsion stability (ESI) and activity indexes (EAI) were measured according to the modified procedure of Ellouze and Pearce et al. [45,46]. Esterified protein solution (1%, *w*/*v*) adjusted to pHs of 5.0, 7.0, and 9.0 was mixed with corn oil at a ratio of 3:1 (*v*/*v*) and homogenized with a high-speed homogenizer (UltraTurrax T25, IKA, Staufen, Germany) at 10,000 r/min for 2 min. After standing for 0 and 10 min, 40 μL aliquots of the emulsion were transferred to 10 mL of 0.1% (*w*/*w*) sodium dodecyl sulfate (SDS). Optical density was recorded at 500 nm using a UV-2450 spectrophotometer (Shimadzu, Kyoto, Japan). EAI and ESI were then calculated using Equations (1) and (2), respectively:

$$\text{EAI}\ (\text{m}^2/\text{g}) = \frac{2 \times 2.303 \times A_1 \times N}{c \times (1 - \varphi) \times 10000} \quad (1)$$

$$\text{ESI}\ (\text{min}) = \frac{A_1}{A_1 - A_2} \times 10 \quad (2)$$

where A_1 is the absorbance of the diluted emulsion immediately after homogenization, N is the dilution factor (250), c is the weight of protein per volume (g/mL), φ is the oil volume fraction in the emulsion (25%), and $A_1 - A_2$ is the difference of the absorbance between time 0 and time 10 min.

3.8. Emulsion Properties

3.8.1. Preparation

Protein solutions (1%, *w/v*) at pHs of 5.0 and 7.0 were used as the water phase, and corn oil was used as the oil phase. The oil and water phase were mixed at a ratio of 2:98 (*v/v*), homogenized at 10,000 r/min for 4 min, and then sonicated by ultrasound probe for 20 min at 500 W using Scientz-II D ultrasound generator (Scientz Biotechnology Co., Ltd., Ningbo, China). During sonication, the temperature of the samples was maintained below 20 °C by utilizing an ice-water bath.

3.8.2. Potential and Particle Size Measurements

The ζ-potentials and droplet sizes of the nano-emulsions with different pHs were determined by a Zetasizer nano-zs90 instrument (Malvern Instruments Ltd., Worcestershire, UK) and a Mastersizer 2000 instrument (Malvern Instruments Ltd., Worcestershire, UK), respectively. The parameter settings of the refractive indices were oil phase (1.46) and water phase (1.33). Samples were diluted 100-fold with deionized water to avoid multiple scattering effects and used for analysis [47].

3.8.3. Microstructure Observation

The microstructure of the emulsions was evaluated by transmission electron microscopy (TEM) (EM 902A, ZEISS, Oberkochen, Germany). The nano-emulsion was diluted to a certain multiple. Then, the diluted nano-emulsion sample was dropped onto a 200-meshed carbon-coated copper grid. The nano-emulsion was stained with 1% (*w/v*) phosphotungstic acid and dried at room temperature for 10 min before being observed using an accelerating voltage of 100 kV.

3.9. Stability of Emulsion

The stability of emulsion was observed by measuring the particle size of emulsion. First, we studied the stability of emulsion under different pH conditions; second, we studied the stability of the emulsion after heat treatment (95 °C, 30 min); third, we studied the stability of the emulsion after freeze-thaw treatment (−20 °C, 24 h); and finally, we studied the storage stability of emulsion (4 °C, 28 days). Please refer to Section 3.8.2 for the determination of emulsion particle size.

3.10. Agar Well Diffusion

According to the modified method of Yuan et al. [48], we evaluated the antibacterial effect of esterified protein emulsion. The 100 mL of inoculum (10^6 CFU/mL) was uniformly spread across the poured nutrient agar medium. Oxford cups of 7.8 mm diameter were filled with 150 μL of the test emulsion and left at 4 °C for 12 h. The petri dish was incubated for 24 h at 36 ± 1 °C. Additionally, the emulsions containing the same concentration of native protein (pH 5.0 and 7.0) were prepared as a control. The zone diameter was measured by a caliper.

3.11. Statistical Analysis

All experiments were performed in triplicate and all data were presented as mean ± standard deviation (SD). Data were subjected to one-way ANOVA and Duncan's significant difference analysis using SPSS software (IBM SPSS statistic 19). The significant level (p) was set as 0.05.

4. Conclusions

In conclusion, the modification by esterification of soy proteins with ethanol changed the structures of the soy proteins, increased the isoelectric point of the soy proteins, and effectively promoted the emulsification of the soy proteins at a low pH range. Under weakly acidic conditions, the esterified soy protein nano-emulsions exhibited excellent storage stability, freeze-thaw stability, and thermal stability, as compared to natural soy protein nano-emulsions. In addition, the modification by esterification also resulted in antibacterial properties in the esterified protein nano-emulsions at a pH of 5.0. Therefore, esterification could broaden the application of the proteins in emulsion delivery systems under low pH conditions.

Author Contributions: T.W.: Conceptualization, Formal analysis, Writing—original draft, Writing—review & editing. K.Y.: Investigation, Methodology, Validation, Writing—review & editing. Y.L.: Data Curation, Software. H.W.: Software, Methodology, Validation. Z.F.: Methodology. H.J.: Writing—review & editing. J.X.: Writing—review & editing. All authors have read and agreed to the published version of the manuscript.

Funding: The Key Research and Development projects of Heilongjiang Province [GA21B001].

Institutional Review Board Statement: Not applicable.

Informed Consent Statement: Not applicable.

Data Availability Statement: Not applicable.

Conflicts of Interest: The authors declare no conflict of interest.

References

1. Gharehbeglou, P.; Jafari, S.M.; Hamishekar, H.; Homayouni, A.; Mirzaei, H. Pectin-whey protein complexes vs. small molecule surfactants for stabilization of double nano-emulsions as novel bioactive delivery systems. *J. Food Eng.* **2019**, *245*, 139–148. [CrossRef]
2. Chang, H.W.; Tan, T.B.; Tan, P.Y.; Abas, F.; Lai, O.M.; Nehdi, I.A.; Tan, C.P. Formation and characterization of thiol-modified fibrillated whey protein isolate solution with enhanced functionalities. *J. Food Eng.* **2017**, *214*, 277–286. [CrossRef]
3. Liu, C.; Wang, Z.J.; Jin, H.; Wang, X.Y.; Gao, Y.; Zhao, Q.S.; Liu, C.H.; Xu, J. Effect of enzymolysis and glycosylation on the curcumin nanoemulsions stabilized by beta-conglycinin: Formation, stability and in vitro digestion. *Int. J. Biol. Macromol.* **2020**, *142*, 658–667. [CrossRef] [PubMed]
4. Fernandez-Avila, C.; Arranz, E.; Guri, A.; Trujillo, A.J.; Corredig, M. Vegetable protein isolate-stabilized emulsions for enhanced delivery of conjugated linoleic acid in Caco-2 cells. *Food Hydrocoll.* **2016**, *55*, 144–154. [CrossRef]
5. Ge, J.; Sun, C.-X.; Mata, A.; Corke, H.; Gan, R.-Y.; Fang, Y. Physicochemical and pH-dependent functional properties of proteins isolated from eight traditional Chinese beans. *Food Hydrocoll.* **2021**, *112*, 106288. [CrossRef]
6. Nesterenko, A.; Alric, I.; Silvestre, F.; Durrieu, V. Comparative study of encapsulation of vitamins with native and modified soy protein. *Food Hydrocoll.* **2014**, *38*, 172–179. [CrossRef]
7. Wang, L.C.; Zhang, H.; Li, H.Y.; Zhang, H.J.; Chi, Y.J.; Xia, N.; Li, Z.S.; Jiang, L.W.; Zhang, X.N.; Rayan, A.M. Fabrication and digestive characteristics of high internal phase Pickering emulsions stabilized by ovalbumin-pectin complexes for improving the stability and bioaccessibility of curcumin. *Food Chem.* **2022**, *389*, 133055. [CrossRef]
8. Mendoza-Sanchez, L.G.; Jimenez-Fernandez, M.; Melgar-Lalanne, G.; Gutiarrez-Lopez, G.F.; Hernandez-Arana, A.; Reyes-Espinosa, F.; Hernandez-Sanchez, H. Chemical lipophilization of bovine alpha-lactalbumin with saturated fatty acyl residues: Effect on structure and functional properties. *J. Agric. Food Chem.* **2019**, *67*, 3256–3265. [CrossRef]
9. Sitohy, M.; Osman, A. Antimicrobial activity of native and esterified legume proteins against Gram-negative and Gram-positive bacteria. *Food Chem.* **2010**, *120*, 66–73. [CrossRef]
10. Sitohy, M.; Chobert, J.M.; Haertlé, T. Improvement of solubility and of emulsifying properties of milk proteins at acid pHs by esterification. *Die Nahr.* **2001**, *45*, 87–93. [CrossRef]

11. Delahaije, R.; Hilgers, R.J.; Wierenga, P.A.; Gruppen, H. Relative contributions of charge and surface coverage on pH-induced flocculation of protein-stabilized emulsions. *Colloid Surf. A* **2017**, *521*, 153–160. [CrossRef]
12. Sitohy, M.; Mahgoub, S.; Osman, A. Controlling psychrotrophic bacteria in raw buffalo milk preserved at 4 degrees C with esterified legume proteins. *LWT* **2011**, *44*, 1697–1702. [CrossRef]
13. Wang, J.; Shi, X.G.; Wang, H.Y.; Xia, X.M.; Wang, K.Y. Effects of esterified lactoferrin and lactoferrin on control of postharvest blue mold of apple fruit and their possible mechanisms of action. *J. Agric. Food Chem.* **2012**, *60*, 6432–6438. [CrossRef]
14. Zhang, D.; Li, H.J.; Wang, Z.F.; Emara, A.M.; Hu, Y.; He, Z.F. Effects of in vitro oxidation on myofibrillar protein charge, aggregation, and structural characteristics. *Food Chem.* **2020**, *332*, 127396. [CrossRef]
15. Wu, C.; Hua, Y.F.; Chen, Y.M.; Kong, X.Z.; Zhang, C.M. Effect of 7S/11S ratio on the network structure of heat-induced soy protein gels: A study of probe release. *RSC Adv.* **2016**, *6*, 101981–101987. [CrossRef]
16. Ulrih, N.P. Analytical techniques for the study of polyphenol-protein interactions. *Crit. Rev. Food. Sci.* **2017**, *57*, 2144–2161. [CrossRef]
17. Li, C.H.; Dai, T.T.; Chen, J.; Li, X.; Li, T.; Liu, C.M.; McClements, D.J. Protein-polyphenol functional ingredients: The foaming properties of lactoferrin are enhanced by forming complexes with procyanidin. *Food Chem.* **2021**, *339*, 128145. [CrossRef]
18. Yang, W.; Xu, C.Q.; Liu, F.G.; Yuan, F.; Gao, Y.X. Native and Thermally modified protein-polyphenol coassemblies: Lactoferrin-based nanoparticles and submicrometer particles as protective vehicles for (-)-epigallocatechin-3-gallate. *J. Agric. Food Chem.* **2014**, *62*, 10816–10827. [CrossRef]
19. Yang, Y.; Wang, Q.; Lei, L.; Li, F.; Zhao, J.; Zhang, Y.; Li, L.; Wang, Q.; Ming, J. Molecular interaction of soybean glycinin and β-conglycinin with (−)-epigallocatechin gallate induced by pH changes. *Food Hydrocoll.* **2020**, *108*, 106010. [CrossRef]
20. Wang, Y.; Wang, D.; Sun, X.S. Effect of esterification on soy protein adhesive performance. In Proceedings of the 2005 ASAE Annual Meeting, Tampa, FL, USA, 17–21 July 2005; pp. 17–20.
21. Chang, C.; Wang, T.R.; Hu, Q.B.; Luo, Y.C. Caseinate-zein-polysaccharide complex nanoparticles as potential oral delivery vehicles for curcumin: Effect of polysaccharide type and chemical cross-linking. *Food Hydrocoll.* **2017**, *72*, 254–262. [CrossRef]
22. Bhattacharya, M.; Jain, N.; Bhasne, K.; Kumari, V.; Mukhopadhyay, S. pH-induced conformational isomerization of bovine serum albumin studied by extrinsic and intrinsic protein fluorescence. *J. Fluoresc.* **2011**, *21*, 1083–1090. [CrossRef]
23. Ijarotimi, O.S.; Malomo, S.A.; Fagbemi, T.N.; Osundahunsi, O.F.; Aluko, R.E. Structural and functional properties of Buchholzia coriacea seed flour and protein concentrate at different pH and protein concentrations. *Food Hydrocoll.* **2018**, *74*, 275–288. [CrossRef]
24. Mattarella, N.L.; Richardson, T. Physicochemical and functional properties of positively charged derivatives of bovine β-lactoglobulin. *J. Agric. Food Chem.* **1983**, *31*, 972–978. [CrossRef]
25. Halpin, M.I.; Richardson, T. Selected functionality changes of β-lactoglobulin upon esterification of side-chain carboxyl groups. *J. Dairy Sci.* **1985**, *68*, 3189–3198. [CrossRef] [PubMed]
26. Li, D.; Zhao, Y.; Wang, X.; Tang, H.; Wu, N.; Wu, F.; Yu, D.; Elfalleh, W. Effects of (+)-catechin on a rice bran protein oil-in-water emulsion: Droplet size, zeta-potential, emulsifying properties, and rheological behavior. *Food Hydrocoll.* **2020**, *98*, 105306. [CrossRef]
27. Zhang, S.; Tian, L.; Yi, J.; Zhu, Z.; Decker, E.A.; McClements, D.J. Mixed plant-based emulsifiers inhibit the oxidation of proteins and lipids in walnut oil-in-water emulsions: Almond protein isolate-camellia saponin. *Food Hydrocoll.* **2020**, *109*, 106136. [CrossRef]
28. Jin, H.; Wang, X.; Chen, Z.; Li, Y.; Liu, C.; Xu, J. Fabrication of β-conglycinin-stabilized nanoemulsions via ultrasound process and influence of SDS and PEG 10,000 co-emulsifiers on the physicochemical properties of nanoemulsions. *Food Res. Int.* **2018**, *106*, 800–808. [CrossRef]
29. Wang, S.N.; Yang, J.J.; Shao, G.Q.; Qu, D.N.; Zhao, H.K.; Zhu, L.J.; Yang, L.N.; Li, R.R.; Li, J.; Liu, H.; et al. Dilatational rheological and nuclear magnetic resonance characterization of oil-water interface: Impact of pH on interaction of soy protein isolated and soy hull polysaccharides. *Food Hydrocoll.* **2020**, *99*, 105366. [CrossRef]
30. Caballero, S.; Davidov-Pardo, G. Comparison of legume and dairy proteins for the impact of Maillard conjugation on nanoemulsion formation, stability, and lutein color retention. *Food Chem.* **2021**, *338*, 128083. [CrossRef]
31. Feng, H.Y.; Jin, H.; Gao, Y.; Yan, S.Q.; Zhang, Y.; Zhao, Q.S.; Xu, J. Effects of freeze-thaw cycles on the structure and emulsifying properties of peanut protein isolates. *Food Chem.* **2020**, *330*, 127215. [CrossRef]
32. Zhu, L.; Yin, P.; Xie, T.; Liu, X.; Yang, L.; Wang, S.; Li, J.; Liu, H. Interaction between soyasaponin and soy β-conglycinin or glycinin: Air-water interfacial behavior and foaming property of their mixtures. *Colloid Surf. B* **2020**, *186*, 110707. [CrossRef]
33. Noshad, M.; Mohebbi, M.; Shahidi, F.; Koocheki, A. Freeze-thaw stability of emulsions with soy protein isolate through interfacial engineering. *Int. J. Refrig.* **2015**, *58*, 253–260. [CrossRef]
34. Zhang, S.; Holmes, M.; Ettelaie, R.; Sarkar, A. Pea protein microgel particles as Pickering stabilisers of oil-in-water emulsions: Responsiveness to pH and ionic strength. *Food Hydrocoll.* **2020**, *102*, 105583. [CrossRef]
35. Sitohy, M.; Mahgoub, S.; Osman, A.; El-Masry, R.; Al-Gaby, A. Extent and mode of action of cationic legume proteins against listeria monocytogenes and salmonella enteritidis. *Probiotics Antimicrob. Proteins* **2013**, *5*, 195–205. [CrossRef]
36. Shao, Z.; Yang, Y.; Fang, S.; Li, Y.; Chen, J.; Meng, Y. Mechanism of the antimicrobial activity of whey protein-epsilon-polylysine complexes against Escherichia coli and its application in sauced duck products. *Int. J. Food Microbiol.* **2020**, *328*, 108663. [CrossRef]

37. Abdel-Shafi, S.; Al-Mohammadi, A.-R.; Osman, A.; Enan, G.; Abdel-Hameid, S.; Sitohy, M. Characterization and antibacterial activity of 7S and 11S globulins isolated from cowpea seed protein. *Molecules* **2019**, *24*, 1082. [CrossRef]
38. Ziani, K.; Chang, Y.; McLandsborough, L.; McClements, D.J. Influence of surfactant charge on antimicrobial efficacy of surfactant-stabilized thyme oil nanoemulsions. *J. Agric. Food Chem.* **2011**, *59*, 6247–6255. [CrossRef] [PubMed]
39. Wang, Y.T.; Wang, Z.J.; Handa, C.L.; Xu, J. Effects of ultrasound pre-treatment on the structure of β-conglycinin and glycinin and the antioxidant activity of their hydrolysates. *Food Chem.* **2017**, *218*, 165–172. [CrossRef]
40. Li, Y.Y.; Jin, H.; Sun, X.T.; Sun, J.Y.; Liu, C.; Liu, C.H.; Xu, J. Physicochemical properties and storage stability of food protein-stabilized nanoemulsions. *Nanomaterials* **2019**, *9*, 25. [CrossRef]
41. Xu, J.; Chen, Z.J.; Han, D.; Li, Y.Y.; Sun, X.T.; Wang, Z.J.; Jin, H. Structural and functional properties changes of-conglycinin exposed to hydroxyl radical-generating systems. *Molecules* **2017**, *22*, 1893. [CrossRef]
42. Dong, D.; Li, X.F.; Hua, Y.F.; Chen, Y.M.; Kong, X.Z.; Zhang, C.M.; Wang, Q. Mutual titration of soy proteins and gum arabic and the complexing behavior studied by isothermal titration calorimetry, turbidity and ternary phase boundaries. *Food Hydrocoll.* **2015**, *46*, 28–36. [CrossRef]
43. Feng, H.Y.; Jin, H.; Gao, Y.; Zhu, X.Q.; Zhao, Q.S.; Liu, C.H.; Xu, J. The effect of (-)-epigallocatechin-3-gallate non-covalent interaction with the glycosylated protein on the emulsion property. *Polymers* **2019**, *11*, 1688. [CrossRef]
44. Li, G.; Xu, J.W.; Wang, H.W.; Jiang, L.Z.; Wang, H.; Zhang, Y.; Jin, H.; Fan, Z.J.; Xu, J.; Zhao, Q.S. Physicochemical antioxidative and emulsifying properties of soybean protein hydrolysates obtained with dissimilar hybrid nanoflowers. *Foods* **2022**, *11*, 3409. [CrossRef]
45. Pearce, K.N.; Kinsella, J.E. Emulsifying properties of proteins: Evaluation of a turbidimetric technique. *J. Agric. Food Chem.* **1978**, *26*, 716–723. [CrossRef]
46. Ellouze, M.; Lajnaf, R.; Zouari, A.; Attia, H.; Ayadi, M.A.; Vial, C. Camel alpha-lactalbumin at the oil-water interface: Effect of protein concentration and pH change on surface characteristics and emulsifying properties. *Colloid Surf. B* **2020**, *189*, 110654. [CrossRef]
47. Xu, J.; Teng, F.; Wang, B.Q.; Ruan, X.X.; Ma, Y.F.; Zhang, D.Y.; Zhang, Y.; Fan, Z.J.; Jin, H. Gel property of soy protein emulsion gel: Impact of combined microwave pretreatment and covalent binding of polyphenols by alkaline method. *Molecules* **2022**, *27*, 3458. [CrossRef]
48. Yuan, Y.; Wan, Z.L.; Yin, S.W.; Yang, X.Q. Stability and antimicrobial property of soy protein/chitosan mixed emulsion at acidic condition. *Food Funct.* **2013**, *4*, 1394–1401. [CrossRef] [PubMed]

molecules

MDPI

Article

Identification and Structure of Epitopes on Cashew Allergens Ana o 2 and Ana o 3 Using Phage Display

Ailin Zhang [1], HuiJuan Zhao [1], Shaohan Pei [1], Yue Chi [1], Xiuhua Fan [1] and Jianqing Liu [2,*]

[1] College of Food Science and Biotechnology, Tianjin Agricultural University, Tianjin 300384, China
[2] Baotou Teachers College, Inner Mongolia University of Science and Technology, Baotou 014030, China
* Correspondence: 60950@bttc.edu.cn

Abstract: Background: Cashew (*Anacardium occidentale* L.) is a commercially important plant. Cashew nuts are a popular food source that belong to the tree nut family. Tree nuts are one of the eight major food allergens identified by the Food and Drug Administration in the USA. Allergies to cashew nuts cause severe and systemic immune reactions. Tree nut allergies are frequently fatal and are becoming more common. Aim: We aimed to identify the key allergenic epitopes of cashew nut proteins by correlating the phage display epitope prediction results with bioinformatics analysis. Design: We predicted and experimentally confirmed cashew nut allergen antigenic peptides, which we named Ana o 2 (cupin superfamily) and Ana o 3 (prolamin superfamily). The Ana o 2 and Ana o 3 epitopes were predicted using DNAstar and PyMoL (incorporated in the Swiss-model package). The predicted weak and strong epitopes were synthesized as peptides. The related phage library was built. The peptides were also tested using phage display technology. The expressed antigens were tested and confirmed using microtiter plates coated with pooled human sera from patients with cashew nut allergies or healthy controls. Results: The Ana o 2 epitopes were represented by four linear peptides, with the epitopes corresponding to amino acids 108–111, 113–119, 181–186, and 218–224. Furthermore, the identified Ana o 3 epitopes corresponding to amino acids 10–24, 13–27, 39–49, 66–70, 101–106, 107–114, and 115–122 were also screened out and chosen as the key allergenic epitopes. Discussion: The Ana o 3 epitopes accounted for more than 40% of the total amino acid sequence of the protein; thus, Ana o 3 is potentially more allergenic than Ana o 2. Conclusions. The bioinformatic epitope prediction produced subpar results in this study. Furthermore, the phage display method was extremely effective in identifying the allergenic epitopes of cashew nut proteins. The key allergenic epitopes were chosen, providing important information for the study of cashew nut allergens.

Keywords: cashew nuts; allergens; phage display technology; epitope prediction

Citation: Zhang, A.; Zhao, H.; Pei, S.; Chi, Y.; Fan, X.; Liu, J. Identification and Structure of Epitopes on Cashew Allergens Ana o 2 and Ana o 3 Using Phage Display. *Molecules* **2023**, *28*, 1880. https://doi.org/10.3390/molecules28041880

Academic Editor: Carmen Cuadrado

Received: 16 January 2023
Revised: 3 February 2023
Accepted: 6 February 2023
Published: 16 February 2023

1. Popular Scientific Summary

Bioinformatic analyses are frequently used to predict protein properties related to antigenicity. We used two antigenic proteins present in cashew nuts to compare the results of the bioinformatics-based antigenicity predictions with those of the experimental epitope mapping. Several amino acid oligopeptides, representing the predicted and unpredicted epitopes, were expressed using phage display and screened with the pooled human sera of patients with a cashew nut allergy. The findings revealed that the computer predictions did not correlate very well with the presence of linear epitopes.

2. Introduction

The cashew plant (*Anacardium occidentale* L.) is a member of the Anacardiaceae family, which includes nine Anacardium species [1]. Cashew nuts are popular as snacks around the world, and their consumption has increased in recent years. However, some people are allergic to cashew nuts, so eating them poses a significant health risk to this population.

With the increasing awareness of food allergies, a cashew nut allergy has become an important public safety and health problem [2]. Food allergy is a global public health problem which has increased significantly in the past decade, affecting the quality of life of consumers and increasing the demand for health service resources [3]. Food allergies affect approximately 2–4% of the population, peaking at 6–8% at the age of one year and gradually declining to 2–4% in older children and adults [4].

Every antigen has its own antigenic determinant or epitope. Epitopes are kinds of specific locations on antigens that could be recognized by immune cells [5]. An epitope typically consists of several amino acid (aa) residues, often (but not always) aligned along the peptide chain, which comes into physical contact with the lymphocytes during an immune reaction. Antibodies can specifically bind with epitopes. Because the binding sites between antibodies and antigens are relatively small, usually, the three-dimensional size of an epitope is only approximately 3 nm $\times$ 1.5 nm $\times$ 0.7 nm; the smaller size of the epitope typically allows about 5 to 7 aa or monosaccharide residues, and the epitope seldom consists of more than 20 aa [6,7].

Until now, several methods have been employed to study the epitopes in macromolecules. Bioinformatic analysis and the use of specific databases allow for a rapid and easy epitope prediction. For instance, Shan et al. successfully predicted the epitope segment of the β-soya globulin α subunit by using three different bioinformatics tools, DNAStar, the BepiPred 1.0, and the SOPMA web server [8]. Hu et al. predicted the dominant epitope in the peanut allergen by using SOPMA software. In contrast, the T-cell epitopes in peanut allergen Ara h 1, together with walnut allergen Jug r 2 and cashew allergen Ana o 1, were identified using various bioinformatics tools [9–12]. However, these methods are not accurate enough to predict such epitopes, and they need to be verified using in vitro assays, such as the highly powerful phage display method.

Phage display, an experimental method to screen peptides with certain properties, was first established by Smith [13] and subsequently used by others for various purposes [14,15]. Because of its continuous development, the phage display is currently being applied to several areas, including an antigen epitope study [16,17], molecular interaction research [18], vaccine development [19–24], and disease diagnosis and therapy [25–28].

Food allergy is a public health issue that has grown significantly in the last decade, affecting consumers' quality of life and placing additional strain on healthcare resources. Furthermore, nut allergies can be fatal. The most common tree nut allergies in the USA are cashew nut allergy and walnut allergy, both of which are increasing globally [29]. So far, it has been determined that the condition is primarily caused by three proteins: Ana o 1 [30], Ana o 2 [31], and Ana o 3 [32]. Ana o 1 (a vicilin) and Ana o 2 (a legumin) are both members of the cupin superfamily, whereas Ana o 3 belongs to the prolamin superfamily [33,34]. Cashew has a low Ana o 1 content and a low sensitivity, making it difficult to isolate and obtain [35].

Here, we used bioinformatics tools to predict the epitopes on the Ana o 2 and Ana o 3 proteins and used the phage display method to verify these predictions by epitope identification. Phage libraries were screened using the human sera of patients with a cashew allergy. Our study provides the theoretical foundation for a better characterization of allergens, thereby assisting the development of hypoallergenic foods in the future.

3. Results

3.1. Identification of Cashew Allergens

The allergens in cashews were isolated using Sephadex G-150 column chromatography, followed by SDS-PAGE electrophoresis and Western blot analysis. The Sephadex purification yielded only four proteins in relatively large quantities. Moreover, the IgE Western blotting of the allergens also revealed two major bands, approximately 33 KDa and 17 KDa. The results of this experiment confirmed the previously reported findings of purified Ana o 2 and Ana o 3 being the two main allergens of cashew nuts. We further performed a structural analysis of these purified proteins.

The main advantage of mass spectrometry in the detection of food allergens is the sequence determination of the peptides and proteins by the protein spectrum. Specifically, a protein was cut into small fragments by specific enzymatic or chemical hydrolysis, and the molecular weight of each product was detected using mass spectrometry. The obtained data were applied to the database to obtain the corresponding known protein and to determine the sequence of the protein to be measured, as shown in Figure 1A. The impact of detection during processing can thus be effectively avoided. The Ana o 2 and Ana o 3 allergens were also identified. The mass spectrometric profile data were compared to the UniProt-Anacardium Occidentale (downloaded on 20 April 2019) and NCBI-Anacardium Occidentale databases. A total of 60,842 profiles and 5 proteins were obtained. The number of peptides obtained from UniProt-*Anacardium occidentale* was 147, and from the NCBI-*Anacardium occidentale*, it was 148. The measured fragment sequences were compared with those obtained from the NCBI database, and two of them were identified to match the sequences of Ana o 2 and Ana o 3. The sequences of Ana o 2 and Ana o 3 are as follows:

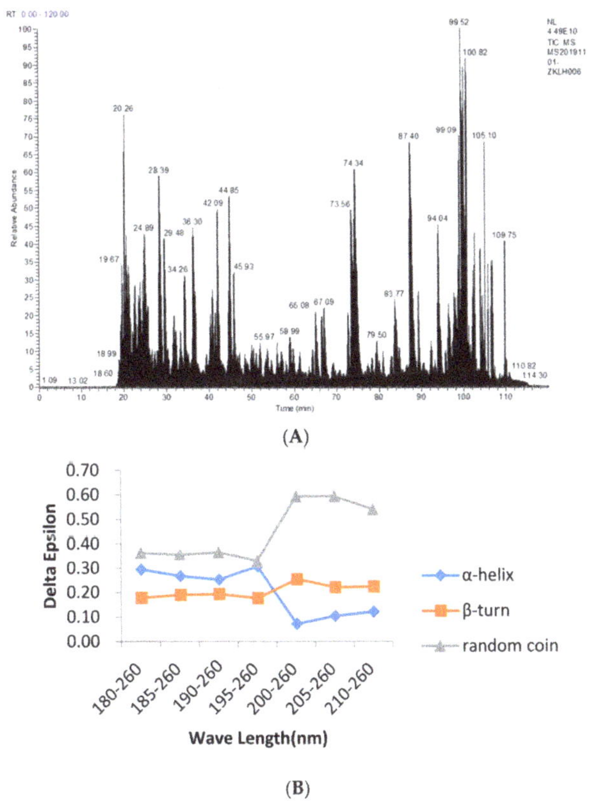

Figure 1. Mass spectrometry analysis of the cashew allergenic proteins (**A**) and the circular dichrogram (**B**) to identify cashew allergenic proteins. Figure (**A**) is from the results of the protein sequencing analysis of cashew allergens. The main purpose of this experiment is to identify two major allergens by Western blotting and obtain the sequences of the two allergens by mass spectrometry sequencing, as it is measured by liquid mass spectrometry. ESI mass spectrometry provides the total ion map. After the NCBI database query, the allergen sequence and the results obtained by mass spectrometry are compared. The secondary structure predicted by the Protein program in DNAstar is compared with the protein structure determined by circular dichroism.

The CD was used to determine the structure of the protein solution, and the CD spectrum in the range of 180–260 nm was provided in the protein secondary structure. From Figure 1B, we found that between 180 and 195 nm, the changes in the α–helix, β–turn, and random coil were relatively stable. The α–helix decreased from 30.70% to 12.20%, the β–turn increased from 17.70% to 22.40%, and the random coil increased from 32.80% to 54.00%, with an overall trend of a decreasing α–helix and an increasing β–turn and random coil. Because of their bulge structures, the β–turn and random coil easily bind to antibodies to form epitopes, whereas the α–helix is not easily deformed and does not easily bind to antibodies to form epitopes, such as tropomyosin, a coiled–coil protein with an α–helix structure that becomes a major allergen from shellfish [36]. Furthermore, a high frequency of β–turn and random coil occurrence was found in the entire amino acid sequence of Ana o 2 and Ana o 3, indicating a high possibility of epitope formation in this region.

3.2. Prediction of the Allergen Secondary Structure with Bioinformatic Analysis

The secondary structures of the allergens Ana o 2 and Ana o 3 were predicted using Protean, a component of the DNAStar software, in conjunction with Swiss-model and PyMoL software; the results are shown in Figure 2 (bioinformatic analysis of Ana o 2 (A) and Ana o 3 (B) features). Those located on the outside of the protein were chosen from among all the identified linear epitopes. Furthermore, those found in the flexible regions and rich in beta–sheets and flexible loops were chosen. Any predicted protruding structure would allow antibodies easy access to the epitope. Therefore, domains that met these requirements were considered potential epitope candidates with a high probability.

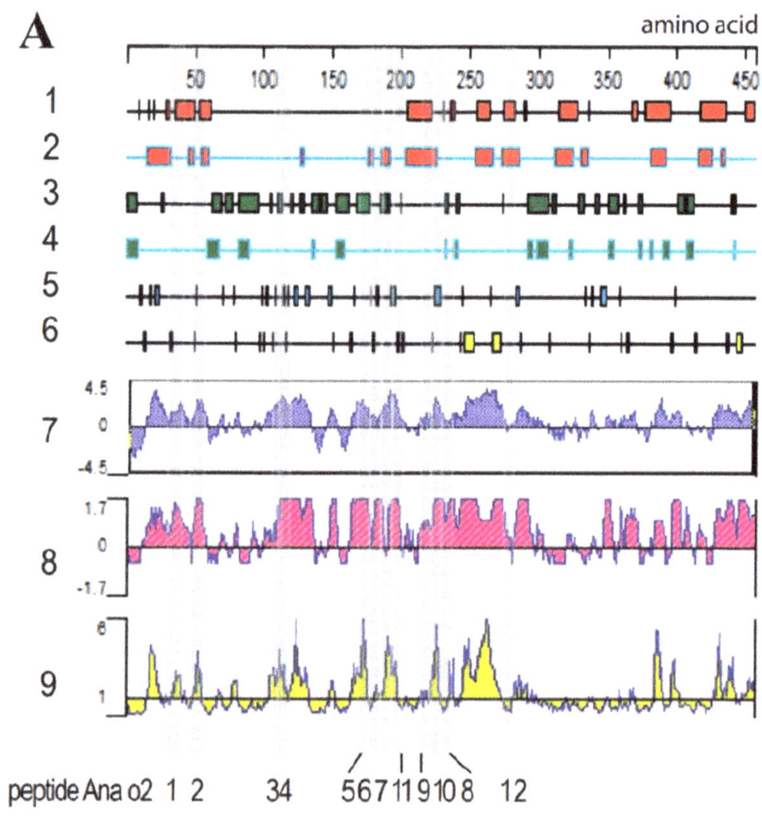

Figure 2. *Cont.*

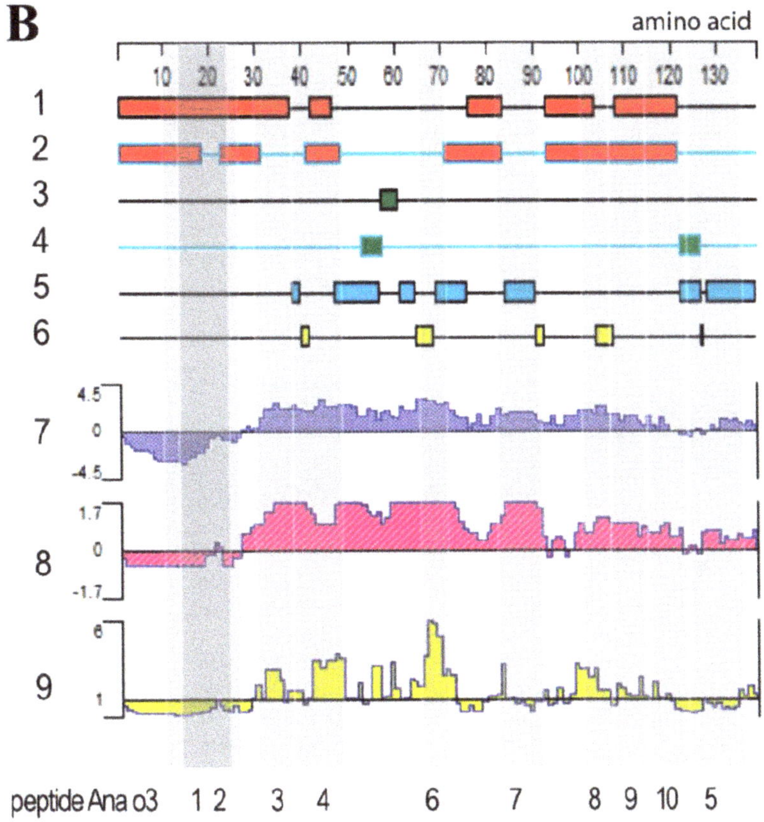

Figure 2. Bioinformatic analysis of the Ana o 2 (**A**) and Ana o 3 (**B**) features. Lanes 1–6: secondary structure analysis revealing alpha regions by using the Garnier–Robson (1) and Chou–Fasman methods (2), beta regions by using the Garnier–Robson (3) and Chou–Fasman methods (4), turn regions by using the Garnier–Robson method (5), and coil regions by using the Garnier–Robson method (6). Further lanes reveal hydrophilicity by using the Kyte–Doolittle method (7), the antigenic index by using the Jameson–Wolf method (8), and the surface probability by using the Emini method (9). Grey shading in both panels corresponds to the produced peptides, whose names are listed below the panels.

The hydrophilicity, surface accessibility, and flexibility of the protein domains are important parameters affecting the antigenic properties. A hydrophilicity above 0, an antigen index above 0, and a surface accessibility of above 1 would, in combination, increase the chance of a protein region representing an epitope. Therefore, a protein domain would have a high surface accessibility as well as the flexibility to fold easily, increasing the exposure at the surface of a three-dimensional protein structure.

The antigen index predicts the likelihood of a protein containing epitopes. As shown in Figure 2B, at least six distinct regions and a large continuous portion of Ana o 2 and Ana o 3 had a high antigen index. To put such predictive indices to the test, we created epitopes with varying lengths and scores based on the antigenicity, hydrophilicity, and surface probability, as shown in Figure 2. Ana o 2's peptides were shorter than those of Ana o 3. As the controls for Ana o 3, two overlapping peptides (Ana o 3–1 and Ana o 3–2) located in regions with negative hydrophilicity and antigenic index values were used. The amino acid sequences were converted to nucleotide sequences through the EMBOSS Backtranseq software (https://www.ebi.ac.uk/Tools/st/emboss_backtranseq/, accessed on 23 Jun 2021); the results are summarized in Table 1.

Table 1. Peptide sequences of the predicted epitopes and the corresponding nucleotide sequences.

Name	Amino Acid Position	Amino Acid Sequence	Nucleotide Sequence
Ana o 2–1	31–41	DALEPDNRVEY	GACGCCCTGGAGCCCGACAACAGGGTGGAGTAC
Ana o 2–2	47–57	EAWDPNHEQFR	GAGGCCTGGGACCCCAACCACGAGCAGTTCAGG
Ana o 2–3	109–113	QQGRQ	CAGCAGGGCAGGCAG
Ana o 2–4	114–120	QGQSGRF	CAGGGCCAGAGCGGCAGGTTC
Ana o 2–5	171–182	FHLAGNP	TTCCACCTGGCCGGCAACCCC
Ana o 2–6	182–187	PKDVFQ	CCCAAGGACGTGTTCCAG
Ana o 2–7	188–194	QQQQHQS	CAGCAGCAGCAGCACCAGAGC
Ana o 2–11	195–203	RGRNLFSGF	AGGGGCAGGAACCTGTTCAGCGGCTTC
Ana o 2–9	219–225	IKQLKSE	ATCAAGCAGCTGAAGAGCGAG
Ana o 2–10	226–232	DNRGGIV	GACAACAGGGGCGGCATCGTG
Ana o 2–8	233–239	KVKDDEL	AAGGTGAAGGACGACGAGCTG
Ana o 2–12	284–288	ENTND	GAGAACACCAACGAC
Ana o 3–1	10–24	AFAVLLLVANASIYR	GCCTTCGCCGTGCTGCTGCT GGTGCCAACGCCAGCATCTACAGG
Ana o 3–2	13–27	VLLLVANASIYRAIV	GTGCTGCTGCTGGTGGCCAA CGCCAGCATCTACAGGGCCATCGTG
Ana o 3–3	30–39	EEDSGREQSC	GAGGAGGACAGCGGCAGGGAGCAGAGCTGC
Ana o 3–4	39–48	QRQFEEQQR	CAGAGGCAGTTCGAGGAGCAGCAGAGG
Ana o 3–6	66–71	YNQRQE	TACAACCAGAGGCAGGAG
Ana o 3–7	83–92	VDRRCRCQNL	GTGGACAGGAGGTGCAGGTGCCAGAACCTG
Ana o 3–8	101–107	QQEQIKG	CAGCAGGAGCAGATCAAGGGC
Ana o 3–9	108–115	EEVRELYE	GAGGAGGTGAGGGAGCTGTACGAG
Ana o 3–10	116–123	TASELPRI	ACCGCCAGCGAGCTGCCCAGGATC
Ana o 3–5	124–138	CSISPSQGCQFQSSY	TGCAGCATCAGCCCCAGCCA GGGCTGCCAGTTCCAGAGCAGCTAC

3.3. Cloning of the Epitope Libraries

Oligonucleotides representing the predicted epitope sequences were cloned into the plasmid pCANTAB5E and transformed into the *E. coli* host DH5alpha. After the insert length was confirmed (Figure 3), the plasmids were electroporated into TGI cells. Three colonies were selected from each transformation plate and sent out for sequencing. After the confirmation of the correct inserts, they were used for phage display screening.

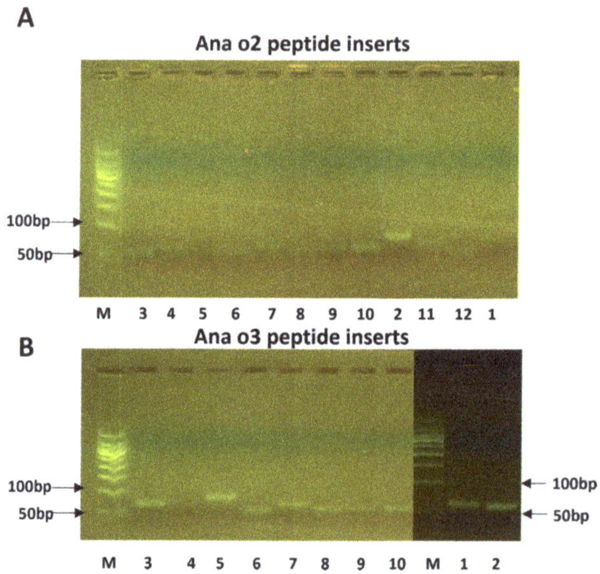

Figure 3. The cloned peptide inserts for Ana o 2 (**A**) and Ana o 3 (**B**). The numbers below the lanes correspond to the peptide numbers provided in Table 1. Lane M represents a size marker.

3.4. Phage Display Screening

Human serum derived from patients with a cashew nut allergy was used to screen the phage display library. This resulted in the identification of major epitopes whose OD_{450} readings are listed in Table 2. Positive signals were obtained for the peptides Ana o 2–3, Ana o 2–4, and Ana o 2–6, as well as a weak positive signal for Ana o 2–9. Despite having a high antigenicity, hydrophilicity, and surface probability, as well as the predicted coil and turn structure, the peptide Ana o 2–2 was not recognized by the pooled antisera. Apart from the presence of a turn but no coil motive, Ana o 2–3 exhibited good but not maximum scores for hydrophilicity, antigenicity, and surface exposure, and it was antigenic. The strongest signal was produced by Ana o 2–4, a peptide of seven amino acids (Table 1), which is located within a stretch with high hydrophilicity (though not producing top scores) and produced maximum scores of antigenicity, combined with a predicted turn and coil region (Figure 3A). Although Ana o 2–5 also contained predicted coil and turn regions together with high predicted hydrophilicity and maximum antigenicity scores, it did not produce positive results in an ELISA. Ana o 2–6 combined maximum antigenicity scores with the presence of a turn but not a coil region; its hydrophilicity was relatively low, and the surface exposure index was even lower. Nevertheless, Ana o 2–6 was recognized as an epitope by the pooled human serum. The weak but positive ELISA results from Ana o 2–9 were unexpected considering the obtained prediction scores.

Table 2. Phage-ELISA results and the predicted antigenicity, hydrophilicity, and surface probability of the peptides.

Peptide	$^{\Delta}$ELISA Value	Hydrophilicity Index [1]	Antigenicity Index [2]	Surface Probability [3]
Ana o 2–1	−0.042	2.0	1.7	2.0
Ana o 2–2	−0.023	2.8	1.7	3.5
Ana o 2–3	**1.558**	2.0	0.8	2.3
Ana o 2–4	**3.532**	3.0	1.7	3.8
Ana o 2–5	−0.044	2.8	1.7	6.0
Ana o 2–6	**1.835**	1.0	1.7	1.0
Ana o 2–7	−0.023	3.8	0.8	4.8
Ana o 2–8	0.010	0.5	1.6	−1.0
Ana o 2–9	**0.292**	1.2	0.6	0.8
Ana o 2–10	−0.031	2.8	1.7	5.8
Ana o 2–11	−0.105	2.0	1.7	2.8
Ana o 2–12	−0.024	0.4	0.4	−1.0
Ana o 3–1	**0.152**	−3.0	−0.6	−1.0
Ana o 3–2	**0.513**	−3.0	−0.6	−1.0
Ana o 3–3	−0.024	2.8	1.7	2.5
Ana o 3–4	**0.66**	3.2	1.7	4.0
Ana o 3–5	−0.106	1.1	0.6	1.0
Ana o 3–6	**0.216**	3.2	1.7	6.0
Ana o 3–7	−0.099	2.0	1.7	4.0
Ana o 3–8	**0.305**	2.3	1.1	3.5
Ana o 3–9	**0.227**	1.5	0.9	3.0
Ana o 3–10	**0.358**	1.3	0.9	2.2

[1]. Peak values correspond to lane 7 of Figure 3. [2]. Peak values correspond to lane 8 of Figure 3. [3]. Peak values correspond to lane 9 of Figure 3. $^{\Delta}$ELISA values are calculated by the peptide sample OD value minus the OD value of the negative control with those above 0 written in bold.

The Ana o 3 peptides produced weaker ELISA signals than the Ana o 2 peptides. Ana o 3–10 produced the strongest ELISA signal, despite having lower hydrophilicity and surface probability scores than Ana o 3–3, which did not represent an epitope. The two overlapping peptides, Ana o 3–1 and Ana o 3–2, were detected by the human antiserum and scored negative for the hydrophilicity, antigenicity, and surface probability (Figure 3B,

Table 2). Therefore, these findings indicated that these bioinformatics tools performed poorly in predicting epitopes, at least for these two proteins.

4. Discussion

We compared the Ana o 2 and Ana o 3 epitopes predicted by in silico analysis to the epitopes identified by phage display technology. The study's goal was to examine the allergen antigenic peptides. It is important to find IgE antibody samples and purify the allergen−specific IgE for the phage. In our experiment, we used an indirect ELISA with a mouse anti−human IgE monoclonal antibody−coated to capture human IgE in pooled samples, avoiding cross−reactivity with other pooled samples or binding with epitopes. We also detected and discovered that the sIgE value (3.2) of the pooled samples was much greater than the 0.35 KUA/L in the patient group and less than 0.35 KUA/L in the healthy control group (0.08). All of this resulted in the important epitopes for the allergic reaction that we are studying being IgE−binding epitopes rather than others−binding epitopes.

Phage display technology was used in this study to determine the main allergen epitopes, primarily because the traditional enzyme−linked immunosorbent assay would inactivate all of the proteins and viruses, resulting in more nonspecific binding reactions of antigen epitopes and interfering with the experimental results. The phage display technology has a higher activity and less nonspecific binding. Because the synthetic peptide's molecular weight was less than 100 bp, many PCR experiments failed, and the single-strand hybridization complementary method was finally chosen to form double strands [37]. When building the transformation library, the electric transformation method was used instead of chemical transformation because it was more efficient. During the experiment, it was discovered that if the peptide was not purified, it could not connect with the bacteriophage. The analysis could be attributed to the dimer-induced inhibition of its binding. Therefore, gel cutting and purification were performed prior to the transformation in order to remove the impurities, such as dimers from the product and improve the linkage rate [38].

From the study of the bioinformatic analysis of the Ana o 2 (A) and Ana o 3 (B) features, between 180 and 195 nm, the changes in the α−helix, β−turn, and random coil were relatively stable. The α−helix decreased from 30.70% to 12.20%, the β−turn increased from 17.70% to 22.40%, and the random coil increased from 32.80% to 54.00%, with an overall trend of a decreasing α−helix and an increasing β−turn and random coil. Because of their bulge structures, the β−turn and random coil easily bind to antibodies to form epitopes, whereas the α−helix is not easily deformed and does not easily bind to antibodies to form epitopes, such as Tropomyosin, a coiled−coil protein with an α−helix structure that becomes a major allergen from shellfish. Furthermore, a high frequency of β−turn and random coil occurrence was found in the entire amino acid sequence of Ana o 2 and Ana o 3, indicating a high possibility of epitope formation in this region.

A previous study used immunoblotting and 15 aa long peptides [39] to perform a systematic epitope screening on Ana o 2. The study identified seven peptides that bound to human IgE, of which aa 105–119 contained Ana o 2–3. The peptide with aa 113–127, on the other hand, did not produce positive results; it contained Ana o 2–4, which we classified as an epitope. The systematic immunoblot screening performed by Wang et al. did not completely cover our epitope Ana o 2–6, while Ana o 2–9 was missed by the corresponding peptide with aa 217–231. Ana o 3 was subjected to a similar systematic epitope screening, with aa 57–68, 72–83, and 102–113 identified as strongly reactive. Our analysis did not confirm the first two, while the third overlapped with our Ana o 3–8 and Ana o 3–9. Immunoblotting can only detect linear epitopes that are not destroyed by denaturation [40]. Phage display screening, on the other hand, allows for some structural features to be conserved, which improves the sensitivity. The use of pooled sera rather than individual serum may have diluted out some of the minor epitopes. Furthermore, the validation of the predicted epitopes is highly dependent on the patient serum; however, the volume of 10 people's mixed serum samples was sufficient to compensate for this.

Therefore, the fraction of aa producing allergen epitopes was 5.5%. Ana o 3 had a much larger fraction because 56 aa produced at least six and possibly seven epitopes, whereas the entire protein was only 138 aa long, resulting in an epitope−producing aa fraction of 40.6%. Ana o 3 is thought to be more likely to cause an allergic reaction than Ana o 2 due to the higher number of identified epitopes and a much smaller size.

5. Materials and Methods

5.1. Preparation of Protein Samples and Patients' Samples

The protein was resolved using Omni PAGE precast gel (4–15%). The protein samples were prepared by mixing with 5X loading buffer (4:1 by volume) and heating for 5 min, followed by cooling at room temperature. The cashew allergen protein was purified using glucan G−150 column chromatography [41]. SDS−PAGE [42] and Western blotting were performed as described in a previous study. Subsequently, 10 μL of the prepared samples were loaded onto the gel (a maximum of 60 μL per well). The gel was run at 150 V until the bromophenol blue dye front reached the gel bottom (40–50 min). Half of the gel was then incubated in a staining solution for 30 min with gentle shaking. For destaining, the gel was incubated with a destaining solution and photographed using a gel imaging system. The other half of the gel was used for Western blotting, and the proteins were visualized by immune detection.

After people with a cashew allergy eat cashews, a special substance (human IgE antibody) appears in their blood, causing hypersensitivity [43]. This study followed all clinical study procedures, as described in the introduction [44]. So, we created two groups: one for patients and one for healthy controls. The patients with cashew nut allergies were identified as the patient group; all of them had a specific IgE antibody value of cashew no less than 0.35 KUA/L with typical and compound clinical allergy symptoms, such as anaphylactic symptoms on the skin or in the respiratory or gastrointestinal areas detected by doctors, and the sera from 10 cases with a cashew nut allergy as the positive serum samples were provided by our collaborators at the Xinzhou People's Hospital. The healthy control group consisted of healthy participants with no obvious history of an allergy, as well as those from Xinzhou People's Hospital. Both groups of participants aged 18–59 years (regardless of gender) were free of other diseases. We used a diagnostic kit to confirm that the patients and control groups did not have any other food allergies when we collected the samples for our group (hob–biotech, Suzhou). The information from these samples is shown in Table 3, which includes sex, age, cashew–specific IgE, cashew reaction history, and diagnosis method. If the SIgE value of the cashew was not less than 0.35 KUA/L, with anaphylactic symptoms on skin or in the respiratory and gastrointestinal areas diagnosed by doctors, it was diagnosed to be that of a cashew allergy patient.

Table 3. The patients' hospital records of cashew allergy.

Group Number	Sample ID	Gender	Age	Specific IgE of Cashew *	Total IgE	Reaction History and Symptoms of Cashew Occurred	Diagnosis Method
Patients Group	SA1	Male	26	0.59	<100	Skin itch, erythema, rhinitis	>0.35 positive
	SA2	Male	22	0.69	<100	Skin itch, erythema, rhinitis	>0.35 positive
	SA3	Female	43	0.47	<100	Skin itch, erythema, rhinitis	>0.35 positive
	SA4	Male	27	0.43	<100	Respiratory tract sneezing, erythema, rhinitis	>0.35 positive
	SA5	Male	35	2.1	<100	Emesis, erythema, rhinitis, skin itch	>0.35 positive
	SA6	Female	19	1.6	<100	Erythema, rhinitis, skin itch	>0.35 positive
	SA7	Female	36	0.39	<100	Skin itch, erythema, rhinitis	>0.35 positive
	SA8	Male	54	0.57	<100	Skin itch, erythema, rhinitis	>0.35 positive
	SA9	Female	42	0.73	<100	Rhinorrhea, erythema, rhinitis, skin itch	>0.35 positive
	SA10	Male	33	7.8	>200	Urticaria, skin itch, erythema, rhinitis	>0.35 positive

Table 3. *Cont.*

Group Number	Sample ID	Gender	Age	Specific IgE of Cashew *	Total IgE	Reaction History and Symptoms of Cashew Occurred	Diagnosis Method
Healthy Control	HC1	Female	23	0.09	<20	Not found	>0.35 positive
	HC2	Female	18	0.13	<20	Not found	>0.35 positive
	HC3	Female	27	0.12	<20	Not found	<0.35 negative
	HC4	Male	34	0.09	<20	Not found	<0.35 negative
	HC5	Male	36	0.05	<20	Not found	<0.35 negative
	HC6	Female	47	0.02	<20	Not found	<0.35 negative
	HC7	Female	42	0.03	<20	Not found	<0.35 negative
	HC8	Male	53	0.02	<20	Not found	<0.35 negative
	HC9	Female	58	0.03	<20	Not found	<0.35 negative
	HC10	Male	32	0.08	<20	Not found	<0.35 negative

* Here if SIgE value of cashew is not less than 0.35 KUA/L with allergic symptoms in skin, respiratory and gastrointestinal detected by doctors, it will diagnose to be cashew allergy patients.

5.2. Mass Spectrometry Sequencing

Protein samples must be pretreated before the enzymatic hydrolysis of protein solutions can take place. To begin, we took some protein samples and mixed them with 20 μL of 2 M urea. After 30 min, 20 μL of 10 mM dithiothreitol (DTT) (dissolved in 25 mM NH_4HCO_3) were added and incubated at 37 °C in a water bath for 1 h. After that, it was filled with 20 μL of 55 mM indole–3–acetic acid (IAA) (dissolved in 25 mM NH_4HCO_3) and left in the dark for 30 min. The samples were then fully mixed and incubated overnight at 37 °C with 0.2 μg of trypsin. The samples were centrifuged the next day and then desalted using the C18 Zip Tip column. The protein solutions were then centrifuged, vacuum freeze–dried, and kept at −20 °C.

The mass spectrometry analysis was run. The peptide fragment was resolubilized in 20 μL of 2% methanol and 0.1% formic acid. After centrifuging the mixture at 12,000× *g* rpm for 10 min, the supernatant was aspirated for sample loading. Following that, 10 μL of the sample was loaded for 15 min using the sandwich method, with a loading pump flow rate of 350 nL/min and a separation flow rate of 300 nL/min.

Following the mass spectrometry sequencing, the peptide sequencing data were compared to those found in the UniProt-*Anacardium occidentale* database (downloaded on 20 April 2019) and the NCBI–*Anacardium occidentale* database. Maxquant (an information retrieval package) was also used to analyze the proteomic data.

5.3. Determination of Secondary Structures by Using Circular Dichroism (CD)

First, the background was measured without any objects being placed in the sample pool. Following that, the sample solution was injected into the reader's cuvette (the machine: Circular Dichroism Spec, Applied PhotoPhysics, APL). Meanwhile, the samples were placed in a cuvette and detected three times at a 0.5 nm bandwidth, 1.0 nm step, 180–260 nm wavelength, and 0.5 s time-per-point.

5.4. Bioinformatics Analysis

The allergen sequences obtained by the mass spectrometry sequencing were compared with those obtained from the NCBI database (https://www.ncbi.nlm.nih.gov/protein/AAN76862.1?report=genpept, accessed on 8 April 2021). Then, the secondary structures of the predicted allergenic proteins were compared using the Protean program in DNAstar, with the structure and binding of the proteins determined using the CD.

5.5. Epitope Prediction

The aa sequences of allergens Ana o 2 and Ana o 3 were extracted from the NCBI database (accession number AF453947_1 and AAL91665, respectively). The antigenic epitopes of these two proteins were predicted using the default settings of the DNAstar and PyMoL software, both of which are part of the Swiss model (https://swissmodel.expasy.org/interactive, accessed on 8 April 2021), along with predictions of their hydrophilicity, flexibility, and surface accessibility. Jin Wei Zhi Biological Co., Ltd. synthesized the DNA oligonucleotides of the predicted epitopes, with upstream and downstream primers designed to contain the *Xma*I and *Kpn*I restriction sites (forward primer: CAGCCCCG-GTACCCCG, reverse primer: GCGGCCGTCCCCGGG), which added 33 bp to the length of each peptide nucleotide sequence [45].

5.6. Epitope Cloning and Expression

The pCANTAB5E phage display system was used to express the predicted epitopes. To see if the peptides were correctly cloned into the plasmid, the epitopes were cloned and expressed in *E. coli* DH5a cells. The pCANTAB5E plasmid was transformed into *E. coli* DH5α, cultured with ampicillin as the selective antibiotic, and extracted using an alkaline lysis plasmid extraction kit for this purpose (Omega). The plasmid DNA was digested with the restriction enzymes *Kpn*I and *Xma*I (NEB), separated on a 1.5% agarose gel, and purified using a gel recovery kit (Promega). The synthesized primers were used separately to amplify the entire predicted peptide in a reaction volume of 10 μL using Taq polymerase. The PCR conditions underwent pre-denaturation for 1 min and 20 cycles of 1 min at 95 °C, 30 s at 55 °C, and 30 s at 72 °C, followed by a 4 min terminal extension at 72 °C. After two cycles, the two complementary primers were combined and annealed for 1 h at 37 °C. The amplicons were separated on 5% agarose gels before being cut out, purified, and ligated into the pCANTAB5E plasmid DNA. The ligation mixture was transformed into competent *E. coli* DH5α cells, and the DNA extracted from single colonies was sequenced by a third party (Sangon Biotech) [46].

The recombinant plasmids were then isolated, quantified, and electroporated into TGI competent cells; the transformed cells were diluted with super optimal broth with a catabolite repression (SOC) culture medium and cultured overnight at 37 °C in 2× YT/Amp/Glu plates with 2% glucose and 100 μg/mL. The colonies on the plates were sequenced the following day.

5.7. Phage Peptide Expression in Microtiter Plates

A bacterial suspension (50 μL) of each primary library was added to a 96-well plate together with a 200 μL 2XYT/AMP/Glu culture medium and incubated at 37 °C overnight with a 250 rpm rotation. The cultures were stored at −70 °C with an equal volume of 100% glycerol. The subcultures were grown in 96-well plates until an OD_{600} of 0.4–0.6 was obtained. To these exponentially growing cells, the helper phage M13KO7 (NEB) was added at a ratio of 10 or 20:1 (phage to bacteria) and incubated at 37 °C for 30 min. Following centrifugation, the cell pellets were resuspended in 150 μL of 2XTY/Amp/Kan. Because the expression of the recombinant peptides depends on the osmotic expression of the *lacZ* promoter with glucose removed in this step, the bacteria were cultured overnight at 37 °C and 300 rpm. Following centrifugation, 50 μL of the supernatant obtained from each well were used for the phage enzyme-linked immunosorbent assay (ELISA).

5.8. Identification with Phage ELISA Method

The plate was coated with a mouse anti-human IgE monoclonal antibody (purchased from Jackson immunoResearch Laboratories Inc.) and incubated overnight at 4 °C. The coated test plates were washed three times with PBST the next day. These plates and an uncoated control plate were blocked for 45 min in PBS with 2% BSA. After washing, the pooled human sera derived from ten individuals with a cashew allergy and the pooled human sera from the control group were added into the wells (100 μL/well) and incubated for 1 h (PBST). After washing the plates, each phage supernatant was used as the antigen (mixed 1:1 with 1% MPBS) and added into the wells. Finally, a 1:4000 dilution of monoclonal mouse anti-M13-HRP (MBL) was used as the secondary antibody. After the final wash, the substrate solutions were added to the plates for 15 min of incubation. After adding the stop solution, the plates were read on a Gen5 plate reader (BioTek Co., Ltd., Santa Clara, CA, USA).

6. Conclusions

Epitopes were identified on two proteins derived from cashew nuts by using phage library expression and screening with human sera obtained from patients with a cashew nut allergy. The bioinformatic predictions correlated poorly with the experimentally verified epitopes. Of the identified epitopes, Ana o 3 is more likely to cause allergic reactions than Ana o 2. Therefore, further research should focus on changing the composition or structure of the identified protein domains to produce low-allergen nuts. We conclude that cashew nut allergen epitopes can be expressed, and their physicochemical properties can be studied, by using the bacteriophage display technology, which provides a theoretical basis for allergen studies in the future.

Author Contributions: Conceptualization, A.Z. and H.Z.; methodology, A.Z.; software, X.F.; validation, H.Z., Y.C. and A.Z.; formal analysis, A.Z.; investigation, J.L.; resources, A.Z.; data curation, H.Z.; writing—original draft preparation, S.P.; writing—review and editing, A.Z.; visualization, H.Z.; supervision, J.L.; project administration, A.Z.; funding acquisition, A.Z. All authors have read and agreed to the published version of the manuscript.

Funding: This study was supported by the Open Project Program of State Key Laboratory of Food Nutrition and Safety, Tianjin University of Science and Technology (No. SKLFNS-KF-201804), the TianJin Key Research and Development Program (No. 20YFZCSN00300), and the National Natural Science Foundation of China (No. 31601421, 81760590).

Institutional Review Board Statement: Not applicable.

Informed Consent Statement: Not applicable.

Data Availability Statement: Not applicable.

Acknowledgments: We thank all the colleagues who contributed to this research.

Conflicts of Interest: The authors declare that they have no conflict of interest.

Abbreviations

aa	Amino acids.
SDS-PAGE	Sodium dodecyl sulfate polyacrylamide gel electrophoresis.
WB	Western blotting.
NCBI	National Center of Biotechnology Information.
ELISA	Enzyme—linked immunosorbent assay.

References

1. GRIN-USDA. *Germplasm*; Resources Information Network—United States Department of Agriculture: Beltsville, MD, USA, 2020. Available online: http://www.ars-grin.gov/ (accessed on 12 January 2016).
2. Mendes, C.; Costa, J.; Vicente, A.A.; Oliveira, M.B.P.; Mafra, I. Cashew Nut Allergy: Clinical Relevance and Allergen Characterisation. *Clin. Rev. Allergy Immunol.* **2019**, *57*, 1–22. [CrossRef] [PubMed]
3. De la Cruz, S.; López-Calleja, I.; Martín, R.; González, I.; Alcocer, M.; García, T. Recent Advances in the Detection of Allergens in Foods. *Methods Mol. Biol.* **2017**, *1592*, 263–295. [PubMed]
4. Madsen, C.B.; Hattersley, S.; Buck, J.; Gendel, S.M.; Houben, G.F.; Hourihane, J.O.; Mackie, A.; Mills, E.N.; Nørhede, P.; Taylor, S.L.; et al. Approaches to risk assessment in food allergy: Report from a workshop developing a framework for assessing the risk from allergenic foods. *Food Chem. Toxicol.* **2009**, *47*, 480–489. [CrossRef] [PubMed]
5. Barlow, D.J.; Edward, M.S.; Thornton, J.M. Continuous and discontinuous protein antigenic determinants. *Nature* **1986**, *322*, 747–748. [CrossRef]
6. Pardoll, D.M.; Topalian, S.L. The role of CD4+ T cell responses in antitumor immunity. *Curr. Opin. Immunol.* **1998**, *10*, 588–594. [CrossRef]
7. Ramana, J.; Mehla, K. Immunoinformatics and Epitope Prediction. *Methods. Mol. Biol.* **2020**, *2131*, 155–171.
8. Xiao, H.D. *Prediction and Processing of Allergenic Epitopes of Soybean and Milk on Their Antigenicity and Structure*; Jiangnan University: Wuxi, China, 2013.
9. Hu, C.Q.; Gao, J.Y.; Luo, C.P.; Chen, H.B. Prediction of Secondary Structure and B Cell Epitopes of Peanut Allergen Ara h 2.02. *Food Sci.* **2009**, *30*, 13–15.
10. Kulis, M.; Pons, L.; Burks, A.W. In vivo and T cell cross-reactivity between walnut, cashew and peanut. *Int. Arch. Allergy Immunol.* **2009**, *148*, 109–117. [CrossRef]

11. DeLong, J.H.; Simpson, K.H.; Wambre, E.; James, E.A.; Robinson, D.; Kwok, W.W. Ara h 1-reactive T cells in individuals with peanut allergy. *J. Allergy Clin. Immunol.* **2011**, *127*, 1211–1218.e3. [CrossRef]
12. Archila, L.D.; Jeong, D.; Pascal, M.; Bartra, J.; Juan, M.; Robinson, D.; Farrington, M.L.; Kwok, W.W. Jug r 2–reactive $CD_4(+)$ T cells have a dominant immune role in walnut allergy. *J. Allergy Clin. Immunol.* **2015**, *136*, 983–992.e7. [CrossRef]
13. Smith, G.P. Filamentous fusion phage: Novel expression vectors that display cloned antigens on the virion surface. *Science* **1985**, *228*, 1315–1317. [CrossRef]
14. Clackson, T.; Wells, J.A. In vitro selection from protein and peptide libraries. *Trends Biotechnol.* **1994**, *12*, 173–184. [CrossRef]
15. Lowman, H.B. Bacteriophage display and discovery of peptide leads for drug development. *Annu. Rev. Biophys. Biomol. Struct.* **1997**, *26*, 401–424. [CrossRef]
16. Smith, G.P.; Petrenko, V.A. Phage Display. *Chem. Rev.* **1997**, *97*, 391–410. [CrossRef]
17. Sidhu, S.S. Phage display in pharmaceutical biotechnology. *Curr. Opin. Biotechnol.* **2000**, *11*, 610–616. [CrossRef]
18. Griffiths, A.D.; Williams, S.C.; Hartley, O.; Tomlinson, I.M.; Waterhouse, P.; Crosby, W.L.; Kontermann, R.E.; Jones, P.T.; Low, N.M.; Allison, T.A.; et al. Isolation of high affinity human antibodies directly from large synthetic repertoires. *EMBO J.* **1994**, *13*, 3245–3260. [CrossRef]
19. de Haard, H.J.; van Neer, N.; Reurs, A.; Hufton, S.E.; Roovers, R.C.; Henderikx, P.; de Brume, A.P.; Arends, J.W.; Hoogenboom, H.R. A large non immunized hurnan Fab fragment phage library that permits rapid isolation and kinetic analysis of high afinity antibodies. *J. Biol. Chem.* **1999**, *274*, 18218. [CrossRef]
20. Folgori, A.; Tafi, R.; Meola, A.; Felici, F.; Galfre, G.; Cortese, R.; Monaci, P.; Nicosia, A. A general Strategy to identify mimotopes of pathological antigens using only random peptide libraries and human sera. *EMBO J.* **1994**, *13*, 2236–2243. [CrossRef]
21. Kouzmitcheva, G.A.; Petrenko, V.A.; Smith, G.P. ldentifying diagnostic peptides for lyme disease through epitope discovery. *Clin. Diagn. Lab. Immunol.* **2001**, *8*, 150–160. [CrossRef]
22. Rullas, J.; Dhar, N.; McKinney, J.D.; Garcia-Perez, A.; Lelievre, J.; Diacon, A.H.; Hugonnet, J.E.; Arthur, M.; Angulo-Barturen, I.; Barros-Aguire, D.; et al. Combinations of beta-Lactam antibiotics currently in clinical trials are efficacious in a DHP-I-deficient mouse model of tuberculosis Infection. *Antimicrob. Agents Chemother.* **2015**, *59*, 4997–4999. [CrossRef]
23. Dhar, N.; Dubee, V.; Balell, L.; Cuinet, G.; Hugonnet, J.E.; Signorino-Gelo, F.; Barros, D.; Arthur, M.; McKinney, J.D. Rapid cytolysis of Myeobacterium tuberculosis by faropenem, an orally bioavailable beta-lactam antibiotic. *Antimicrob. Agents Chemother.* **2015**, *59*, 1308–1319. [CrossRef] [PubMed]
24. Lavollay, M.; Arthur, M.; Fourgeaud, M.; Dubost, L.; Marie, A.; Vezinis, N.; Blanol, D.; Gutmann, L.; Mainardi, J.L. The peptidoglycan of stationary-phase Mycobacterium tuberculosis predominantly contains cross-inks generated by L,D-transpeptidation. *J. Bacteriol.* **2008**, *190*, 4360–4366. [CrossRef] [PubMed]
25. Vollmer, W.; Blanot, D.; de Pedro, M.A. Peptidoglycan structure and architecture. *FEMS Microbiol. Rev.* **2008**, *32*, 149–167. [CrossRef] [PubMed]
26. Bertelli, A.A.; Ferrara, F.; Diana, G.; Fulgenzi, A.; Corsi, M.; Ponti, W.; Ferrero, M.E.; Bertelli, A. Resveratrol, a natural stilbene in grapes and wine, enhances intraphagocytosis in human promonocytes: A co-factor in antiinflammatory and anticancer chemopreventive activity. *Int. J. Tissue React.* **1999**, *21*, 93–104. [PubMed]
27. Gupta, R.S.; Springston, E.E.; Warrier, M.R.; Smith, B.; Kumar, R.; Pongracic, J.; Holl, J.L. The prevalence, severity, and distribution of childhood food allergy in the United States. *Pediatrics* **2011**, *128*, e9–e17. [CrossRef]
28. Dhanapala, P.; De Silva, C.; Doran, T.; Suphioglu, C. Cracking the Egg: An Insight into Egg hypersensitivity. *Mol. Immunol.* **2015**, *66*, 375–383. [CrossRef]
29. Grifiths, A.D.; Duncan, A.R. Strategies for selection of antibodies by phage display. *Curr. Opin. Biotechnol.* **1998**, *9*, 102–108. [CrossRef]
30. Becker, M.; Felsberger, A.; Frenzel, A.; Shattuck, W.M.C.; Dyer, M.; Kügler, J.; Zantow, J.; Mather, T.N.; Hust, M. Application of M_{13} phage display for identifying immunogenic proteins fom tick (*Ixodes scapularis*) saliva. *BMC Biotechnol.* **2015**, *15*, 43. [CrossRef]
31. Sicherer, S.H.; Furlong, T.J.; Muñoz-Furlong, A.; Burks, A.W.; Sampson, H.A. A voluntary registry for peanut and tree nut allergy: Characteristics of the first 5149 registrants. *J. Allergy Clin. Immunol.* **2001**, *108*, 128–132. [CrossRef]
32. Wang, F.; Robotham, J.M.; Teuber, S.S.; Tawde, P.; Sathe, S.K.; Roux, K.H. Ana o 1, a cashew (*Anacardium occidental*) allergen of the vicilin seed storage protein family. *J. Allergy Clin. Immunol.* **2002**, *110*, 160–166. [CrossRef]
33. Wang, F.; Robotham, J.M.; Teuber, S.S.; Sathe, S.K.; Roux, K.H. Ana o 2, a major cashew (*Anacardium occidentale* L.) nut allergen of the legumin family. *Int. Arch. Allergy Immunol.* **2003**, *132*, 27–39. [CrossRef]
34. Robotham, J.M.; Wang, F.; Seamon, V.; Teuber, S.S.; Sathe, S.K.; Sampson, H.A.; Beyer, K.; Seavy, M.; Roux, K.H. Ana o 3, an important cashew nut (*Anacardium occidentale* L.) allergen of the 2S albumin family. *J. Allergy Clin. Immunol.* **2005**, *115*, 1284–1290. [CrossRef]
35. Zaffran, V.D.; Gupta, S.; Sathe, S.K.; Roper, M.G. Effect of deglycosylation on immunoreactivity and in vitro pepsin digestibility of major cashew (*Anacardium occidentale* L.) allergen, Ana o 1. *J. Food Sci.* **2021**, *86*, 1144–1177. [CrossRef]
36. Ochiai, Y.; Ozawa, H. Biochemical and physicochemical characteristics of the major muscle proteins from fish and shellfsh. *Fish. Sci.* **2020**, *86*, 729–740. [CrossRef]
37. Liu, G.M.; Liang, Y.L.; Su, W.J.; Zhang, L.J.; Guo, L.L.; Cao, M.J. Purification and Allergenicity Identification of Carp Parvalbumin. *Food Sci.* **2009**, *30*, 188–191.

38. García, F.; Moneo, I.; Fernández, B.; García-Menaya, J.M.; Blanco, J.; Juste, S.; Gonzalo. Allergy to Anacardiaceae: Description of cashew and pistachio nut allergens. *J. Investig. Allergol. Clin. Immunol.* **2000**, *10*, 173–177.
39. Wang, W.; Han, J.X.; Wu, Y.J.; Yuan, F.; Chen, Y.; Ge, Y.Q. Simultaneous detection of eight food allergens using optical thin-film biosensor chips. *J. Agric. Food Chem.* **2011**, *59*, 6889–6894. [CrossRef]
40. Elizur, A.; Appel, M.Y.; Nachshon, L.; Levy, M.B.; Epstein-Rigbi, N.; Koren, Y.; Holmqvist, M.; Porsch, H.; Lidholm, J.; Goldberg, M.R. Cashew oral immunotherapy for desensitizing cashew-pistachio allergy (NUT CRACKER study). *Allergy* **2022**, *77*, 1863–1872. [CrossRef]
41. Röntynen, P.; Kukkonen, K.; Savinko, T.; Mäkelä, M.J. Optimizing tools for evaluating challenge outcomes in children with cashew nut allergy. *Ann. Allergy Asthma Immunol.* **2022**, *128*, 270–278. [CrossRef]
42. Zhang, A.; Wang, C.; Hu, Y.; Zhou, Z. Isolation of allergens in nuts and analysis on Western Blotting. *Mod. Food Sci. Technol.* **2014**, *30*, 99–104.
43. Zhu, Q.; Wu, X.; Xiao, J.; Lu, J.; Liu, Z.G.; Xu, H. Prediction of the structure and antigenic epitopes of cashew allergen Ana o 2. *J. Food Sci. Biotechnol.* **2012**, *31*, 141–145.
44. Yang, H. *Study on a New Generation of Peptide de Novo Sequencing Based on Mass Spectrometry*; University of Chinese Academy of Sciences: Beijing, China, 2018.
45. Sánchez-Ruano, L.; de la Hoz, B.; Martínez-Botas, J. Clinical utility of microarray B-cell epitope mapping in food allergies: A systematic review. *Pediatr. Allergy Immunolgy* **2020**, *31*, 175–185. [CrossRef] [PubMed]
46. Zhou, X.; Yang, M. *Isolation and Purification of Major Allergens and Identification of Immune Activity*; Zhejiang University: Hangzhou, China, 2015.

MDPI
St. Alban-Anlage 66
4052 Basel
Switzerland
www.mdpi.com

Molecules Editorial Office
E-mail: molecules@mdpi.com
www.mdpi.com/journal/molecules

www.ingramcontent.com/pod-product-compliance
Lightning Source LLC
Chambersburg PA
CBHW062032220926
51850CB00010B/2518
* 9 7 8 3 0 3 6 5 9 8 6 8 0 *